Quick Review Math Handbook

Book 2

congruent angle

intercept

perimeter

percent

surface area

 Glencoe

New York, New York   Columbus, Ohio   Chicago, Illinois   Woodland Hills, California

The *McGraw·Hill* Companies

 Glencoe

Send all inquiries to:
Glencoe/McGraw-Hill
8787 Orion Place
Columbus, OH 43240-4027

ISBN: 978-0-07-891506-2   *(Student Edition)*
MHID: 0-07-891506-6   *(Student Edition)*
ISBN: 978-0-07-891507-9   *(Teacher Wraparound Edition)*
MHID: 0-07-891507-4   *(Teacher Wraparound Edition)*

Printed in the United States of America.

2 3 4 5 6 7 8 9 10 071 17 16 15 14 13 12 11 10 09

Handbook
at a Glance

Handbook
Contents

PART TWO

HotTopics 66

A reference to key topics spread over eight areas of mathematics

1 Numbers and Computation

2 Fractions, Decimals, and Percents

CONTENTS

3 Powers and Roots

4 Data, Statistics, and Probability

4•2 Displaying Data

4•3 Analyzing Data

4•4 Statistics

4•5 Probability

5 Algebra

5.4 Solving Linear Equations

5.5 Ratio and Proportion

5.6 Inequalities

5.7 Graphing on the Coordinate Plane

5.8 Slope and Intercept

6 Geometry

7 Measurement

CONTENTS

8 Tools

PART THREE 3

Handbook
Introduction

Why use this handbook?

You will use this handbook to refresh your memory of mathematics concepts and skills.

What are HotWords, and how do you find them?

HotWords are important mathematical terms. The HotWords section includes a glossary of terms, a collection of common or significant mathematical patterns, and lists of symbols and formulas in alphabetical order. Many entries in the glossary will refer you to chapters and topics in the HotTopics section for more detailed information.

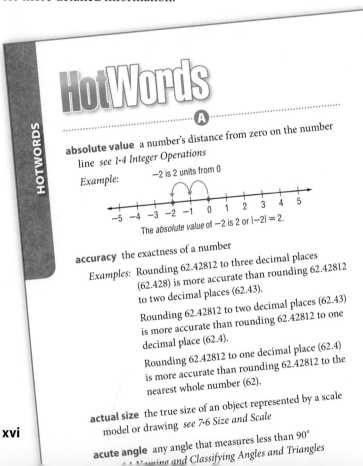

HOTWORDS

HotWords

········· Ⓐ ·········

absolute value a number's distance from zero on the number line *see 1·4 Integer Operations*

Example: −2 is 2 units from 0

The *absolute value* of −2 is 2 or |−2| = 2.

accuracy the exactness of a number

Examples: Rounding 62.42812 to three decimal places (62.428) is more accurate than rounding 62.42812 to two decimal places (62.43).

Rounding 62.42812 to two decimal places (62.43) is more accurate than rounding 62.42812 to one decimal place (62.4).

Rounding 62.42812 to one decimal place (62.4) is more accurate than rounding 62.42812 to the nearest whole number (62).

actual size the true size of an object represented by a scale model or drawing *see 7·6 Size and Scale*

acute angle any angle that measures less than 90°
Naming and Classifying Angles and Triangles

What are HotTopics, and how do you use them?

HotTopics are key concepts that you need to know. The HotTopics section consists of eight chapters. Each chapter has several topics that give you to-the-point explanations of key mathematical concepts. Each topic includes one or more concepts. Each section includes Check It Out exercises, which you can use to check your understanding. At the end of each topic, there is an exercise set.

There are problems and a vocabulary list at the beginning and end of each chapter to help you preview what you know and review what you have learned.

What are HotSolutions?

The HotSolutions section gives you easy-to-locate answers to Check It Out and What Do You Know? problems. The HotSolutions section is at the back of the handbook.

ORDER OF OPERATIONS

1•2

1•2 Order of Ope

Understanding the Order of Op

Solving a problem may involve using more tha
Your answer will depend on the order in which
those operations.

For example, consider the expression $2^2 + 5 \times$ (

$$2^2 + 5 \times 6$$
$$4 + 5 \times 6$$
$$9 \times 6 = \boxed{54}$$

or

$$2^2 + 5 \times 6$$
$$4 + 5 \times 6$$
$$4 + 30 = \boxed{34}$$

The order in which you perform operations makes a difference

To make sure that there is just one answer to a serie
computations, mathematicians have agreed upon ar
which to do the operations.

EXAMPLE   Using the Order of Operations

Simplify $4^2 - 8 \times (6 - 6)$.

$= 4^2 - 8 \times (0)$

$= (16) - 8 \times 0$

$= 16 - 0$

$= 16$

So, $4^2 - 8 \times (6 - 6) = 16$.

- Simplify within parenthe:
- Evaluate the power.
- Multiply or divide from lef
- Add or subtract from left to

Check It Out

Simplify.

angle

line

syr

outcome

parallel

perp

Hot Words

The **HotWords** section includes a glossary of terms, lists of formulas and symbols, and a collection of common or significant mathematical patterns. Many entries in the glossary will refer to chapters and topics in the **HotTopics** section.

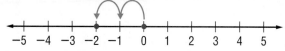

·· Ⓐ ··

✓ **absolute value** a number's distance from zero on the number line *see 1·4 Integer Operations*

Example: −2 is 2 units from 0

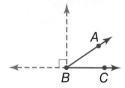

The *absolute value* of −2 is 2 or |−2| = 2.

accuracy the exactness of a number

Examples: Rounding 62.42812 to three decimal places (62.428) is more accurate than rounding 62.42812 to two decimal places (62.43).

Rounding 62.42812 to two decimal places (62.43) is more accurate than rounding 62.42812 to one decimal place (62.4).

Rounding 62.42812 to one decimal place (62.4) is more accurate than rounding 62.42812 to the nearest whole number (62).

actual size the true size of an object represented by a scale model or drawing *see 7·6 Size and Scale*

acute angle any angle that measures less than 90°
see 6·1 Naming and Classifying Angles and Triangles

Example:

∠ABC is an *acute angle*.
0° < m∠ABC < 90°

acute triangle a triangle in which all angles measure less than 90° *see 6·1 Naming and Classifying Angles and Triangles*

Example:

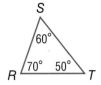

△*RST* is an *acute triangle.*

Addition Property of Equality the mathematical rule that states that if the same number is added to each side of an equation, the expressions remain equal

Example: If $a = b$, then $a + c = b + c$.

additive inverse two integers that are opposite of each other; the sum of any number and its *additive inverse* is zero *see 5·4 Solving Linear Equations*

Example: $(+3) + (-3) = 0$
(-3) is the *additive inverse* of 3.

additive system a mathematical system in which the values of individual symbols are added together to determine the value of a sequence of symbols

Example: The Roman numeral system, which uses symbols such as I, V, D, and M, is a well-known *additive system.*

This is another example of an additive system:
▽▽□
If □ equals 1 and ▽ equals 7,
then ▽▽□ equals $7 + 7 + 1 = 15$.

algebra a branch of mathematics in which symbols are used to represent numbers and express mathematical relationships *see Chapter 5 Algebra*

algorithm a step-by-step procedure for a mathematical operation *see 2·3 Addition and Subtraction of Fractions, 2·4 Multiplication and Division of Fractions, 2·6 Decimal Operations*

altitude the perpendicular distance from a vertex to the opposite side of a figure; *altitude* indicates the height of a figure

Example:

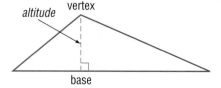

angle two rays that meet at a common endpoint *see 6·1 Naming and Classifying Angles and Triangles*

Example:

∠*ABC* is formed by $\overrightarrow{BA}$ and $\overrightarrow{BC}$.

angle of elevation the angle formed by a horizontal line and an upward line of sight

Example:

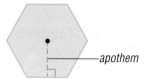

apothem a perpendicular line segment from the center of a regular polygon to one of its sides

Example:

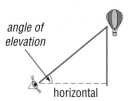

Arabic numerals (or Hindu-Arabic numerals) the number symbols we presently use in our base-ten number system {0, 1, 2, 3, 4, 5, 6, 7, 8, 9}

arc a section of a circle

Example:

$\overarc{QR}$ is an *arc*.

area the measure of the interior region of a two-dimensional figure or the surface of a three-dimensional figure, expressed in square units *see Formulas page 60, 3·1 Powers and Exponents, 6·5 Area, 6·6 Surface Area, 6·8 Circles, 7·3 Area, Volume, and Capacity*

Example:

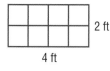

4 ft

area = 8 ft²

arithmetic expression a mathematical relationship expressed as a number, or two or more numbers with operation symbols *see expression*

arithmetic sequence *see Patterns page 63*

Associative Property the mathematical rule that states that the way in which numbers are grouped when they are added or multiplied does not change their sum or product *see 1·1 Properties, 5·2 Simplifying Expressions*

Examples: $(x + y) + z = x + (y + z)$
$x \times (y \times z) = (x \times y) \times z$

average the sum of a set of values divided by the number of values *see 4·4 Statistics*

Example: The *average* of 3, 4, 7, and 10 is
$(3 + 4 + 7 + 10) \div 4 = 6.$

average speed the average rate at which an object moves

axis (pl. *axes*) [1] a reference line by which a point on a coordinate graph may be located; [2] the imaginary line about which an object may be said to be symmetrical (*axis* of symmetry); [3] the line about which an object may revolve (*axis* of rotation) *see 5·7 Graphing on the Coordinate Plane, 6·3 Symmetry and Transformations*

·· **B** ··

bar graph a display of data that uses horizontal or vertical bars to compare quantities *see 4·2 Displaying Data*

base [1] the number used as the factor in exponential form; [2] two parallel congruent faces of a prism or the face opposite the apex of a pyramid or cone; [3] the side perpendicular to the height of a polygon; [4] the number of characters in a number system *see 3·1 Powers and Exponents, 6·5 Surface Area, 6·7 Volume*

base-ten system the number system containing ten single-digit symbols {0, 1, 2, 3, 4, 5, 6, 7, 8, and 9} in which the numeral 10 represents the quantity ten *see 2·5 Naming and Ordering Decimals*

base-two system the number system containing two single-digit symbols {0 and 1} in which 10 represents the quantity two *see binary system*

benchmark a point of reference from which measurements and percents can be estimated *see 2·7 Meaning of Percents*

best chance in a set of values, the event most likely to occur *see 4·5 Probability*

biased sample a sample drawn in such a way that one or more parts of the population are favored over others *see 4·1 Collecting Data*

bimodal distribution a statistical model that has two different peaks of frequency distribution *see 4·3 Analyzing Data*

binary system the base-two number system, in which combinations of the digits 1 and 0 represent different numbers or values

binomial an algebraic expression that has two terms

Examples: $x^2 + y; x + 1; a - 2b$

budget a spending plan based on an estimate of income and expenses

capacity the amount that can be held in a container
see 7·3 Area, Volume, and Capacity

cell a small rectangle in a spreadsheet that stores information; each *cell* can store a label, number, or formula
see 8·3 Spreadsheets

center of the circle the point from which all points on a circle are equidistant *see 6·8 Circles*

chance the probability or likelihood of an occurrence, often expressed as a fraction, decimal, percentage, or ratio
see 4·5 Probability

circle the set of all points in a plane that are equidistant from a fixed point called the center *see 6·8 Circles*

Example:

a *circle*

circle graph (pie chart) a display of statistical data that uses a circle divided into proportionally sized "slices"
see 4·2 Displaying Data

Example: **Favorite Primary Color**

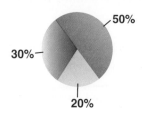

circumference the distance around (perimeter) a circle
see Formulas page 61, 6·8 Circles

classification the grouping of elements into separate classes or sets

coefficient the numerical factor of a term that contains a variable

collinear a set of points that lie on the same line

Example:

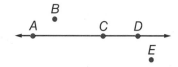

Points *A, C,* and *D* are *collinear.*

columns vertical lists of numbers or terms *see 8·3 Spreadsheets*

combination a selection of elements from a larger set in which the order does not matter *see 4·5 Probability*

Example: 456, 564, and 654 are one *combination* of three digits from 4567.

common denominator a common multiple of the denominators of a group of fractions *see 2·3 Addition and Subtraction of Fractions*

Example: The fractions $\frac{3}{4}$ and $\frac{7}{8}$ have a *common denominator* of 8.

common difference the difference between any two consecutive terms in an arithmetic sequence

common factor a whole number that is a factor of each number in a set of numbers *see 1·3 Factors and Multiples*

Example: 5 is a *common factor* of 10, 15, 25, and 100.

common ratio the ratio of any term in a geometric sequence to the term that precedes it

Commutative Property the mathematical rule that states that the order in which numbers are added or multiplied does not change their sum or product *see 1·1 Properties, 5·2 Simplifying Expressions*

Examples: $x + y = y + x$
$x \times y = y \times x$

compatible numbers two numbers that are easy to add, subtract, multiply, or divide mentally

complementary angles two angles are complementary if the sum of their measures is 90° *see 7·1 Classifying Angles and Triangles*

∠1 and ∠2 are *complementary angles.*

composite number a whole number greater than 1 having more than two factors *see 1·2 Factors and Multiples*

concave polygon a polygon that has an interior angle greater than 180°

Example:

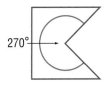

a *concave polygon*

cone a three-dimensional figure consisting of a circular base and one vertex

Example:

vertex

a *cone*

congruent having the same size and shape; the symbol ≅ is used to indicate congruence *see 6·1 Naming and Classifying Angles and Triangles*

Example:

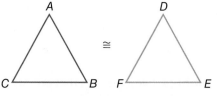

△*ABC* and △*DEF* are *congruent.*

congruent angles angles that have the same measure *see 6·1 Naming and Classifying Angles and Triangles*

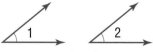

∠1 and ∠2 are *congruent angles.*

conic section the curved shape that results when a conical surface is intersected by a plane

Example:

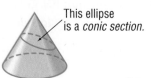

This ellipse is a *conic section.*

continuous data the complete range of values on the number line

Example: The possible sizes of apples are *continuous data.*

convenience sampling a sample obtained by surveying people that are easiest to reach; *convenience sampling* does not represent the entire population; therefore, it is considered biased *see 4·1 Collecting Data*

convex polygon a polygon with all interior angles measuring less than 180° *see 6·2 Polygons and Polyhedrons*

Example:

A regular hexagon is a *convex polygon.*

coordinate any number within a set of numbers that is used to define a point's location on a line, on a surface, or in space *see 1·3 Integer Operations*

coordinate plane a plane in which a horizontal number line and a vertical number line intersect at their zero points *see 5·7 Graphing on the Coordinate Plane, 5·8 Slope and Intercept*

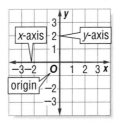

coplanar points or lines lying in the same plane

correlation the way in which a change in one variable corresponds to a change in another *see 4·3 Analyzing Data*

corresponding angles in the figure below, transversal *t* intersects lines ℓ and *m*; ∠1 and ∠5, ∠2 and ∠6, ∠4 and ∠8, and ∠3 and ∠7 are *corresponding angles*; if lines ℓ and *m* are parallel, then these pairs of angles are congruent
see 6·1 Naming and Classifying Angles and Triangles

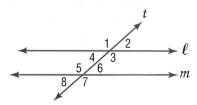

cost an amount paid or required in payment

cost estimate an approximate amount to be paid or to be required in payment

counting numbers the set of positive whole numbers {1, 2, 3, 4 . . .} *see positive integers*

cross product a method used to solve proportions and test whether ratios are equal *see 2·1 Comparing and Ordering Fractions, 5·5 Ratio and Proportion*
Example: if $a \times d = b \times c$

cross section the figure formed by the intersection of a solid and a plane

Example: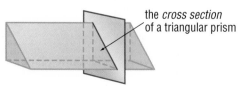
the *cross section* of a triangular prism

cube [1] a solid figure with six congruent square faces; [2] the product of three equal terms *see 3·1 Powers and Exponents, 6·2 Polygons and Polyhedrons, 8·1 Scientific Calculator*

Examples: [1]

a *cube*

[2] $2^3 = 2 \times 2 \times 2 = 8$

cube root a number that when raised to the third power equals a given number *see 8·1 Scientific Calculator*

Example: 2 is the *cube root* of 8.
$$\sqrt[3]{8} = 2$$

cubic centimeter the volume of a cube with edges that are 1 centimeter in length *see 6·7 Volume*

cubic foot the volume of a cube with edges that are 1 foot in length *see 6·7 Volume*

cubic inch the volume of a cube with edges that are 1 inch in length *see 6·7 Volume*

cubic meter the volume of a cube with edges that are 1 meter in length *see 6·7 Volume*

customary system units of measurement used in the United States to measure length in inches, feet, yards, and miles; capacity in cups, pints, quarts, and gallons; weight in ounces, pounds, and tons; and temperature in degrees Fahrenheit *see English system, 7·1 Systems of Measurement*

cylinder a solid shape with parallel circular bases
see 6·6 *Surface Area*

Example:

a *cylinder*

decagon a polygon with ten angles and ten sides
see 6·2 *Polygons and Polyhedrons*

decimal system the most commonly used number system in
which whole numbers and fractions are represented using
base ten *see 2·5 Naming and Ordering Decimals*

Example: Decimal numbers include 1230, 1.23, 0.23, and −13.

degree [1] (algebraic) the exponent of a single variable in a
simple algebraic term; [2] (algebraic) the sum of the exponents
of all the variables in a more complex algebraic term;
[3] (algebraic) the highest degree of any term in a polynomial;
[4] (geometric) a unit of measurement of an angle or arc,
represented by the symbol ° *see 3·1 Powers and Exponents,*
6·1 Naming and Classifying Angles and Triangles, 6·8 Circles,
8·2 Scientific Calculator

Examples: [1] In the term $2x^4y^3z^2$, x has a *degree* of 4, y has a
degree of 3, and z has a *degree* of 2.

[2] The term $2x^4y^3z^2$ as a whole has a *degree* of
$4 + 3 + 2 = 9$.

[3] The equation $x^3 = 3x^2 + x$ is an equation of the
third *degree*.

[4] An acute angle is an angle that measures less
than 90°.

denominator the bottom number in a fraction representing the total number of equal parts in the whole *see 2·1 Fractions and Equivalent Fractions*

Example: In the fraction $\frac{a}{b}$, b is the *denominator*.

dependent events two events in which the outcome of one event is affected by the outcome of another event *see 4·5 Probability*

diagonal a line segment connecting two non-adjacent vertices of a polygon *see 6·2 Polygons and Polyhedrons*

Example:

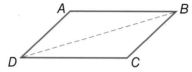

$\overline{BD}$ is a *diagonal* of parallelogram *ABCD*.

diameter a line segment connecting the center of a circle with two points on its perimeter *see 6·8 Circles*

Example:

diameter

difference the result obtained when one number is subtracted from another *see 6·1 Writing Expressions and Equations*

dimension the number of measures needed to describe a figure geometrically

Examples: A point has 0 *dimensions*.
A line or curve has 1 *dimension*.
A plane figure has 2 *dimensions*.
A solid figure has 3 *dimensions*.

direct correlation the relationship between two or more elements that increase and decrease together
see 4·3 Analyzing Data

Example: At an hourly pay rate, an increase in the number of hours worked means an increase in the amount paid, while a decrease in the number of hours worked means a decrease in the amount paid.

discount a deduction made from the regular price of a product or service *see 2·8 Using and Finding Percents*

discrete data only a finite number of values is possible

Example: The number of parts damaged in a shipment is discrete data.

distance the length of the shortest line segment between two points, lines, planes, and so forth *see 7·2 Length and Distance*

Distributive Property the mathematical rule that states that multiplying a sum by a number gives the same result as multiplying each addend by the number and then adding the products *see 1·1 Properties, 5·2 Simplifying Expressions*

Example: $a(b + c) = a \times b + a \times c$

divisible a number is *divisible* by another number if their quotient has no remainder *see 1·2 Factors and Multiples*

division the operation in which a dividend is divided by a divisor to obtain a quotient

Example:

$$12 \div 3 = 4$$

dividend | quotient
divisor

Division Property of Equality the mathematical rule that states that if each side of an equation is divided by the same nonzero number, the two sides remain equal *see 5·4 Solving Linear Equations*

Example: If $a = b$, then $\frac{a}{c} = \frac{b}{c}$.

domain the set of input values in a function

double-bar graph a display of data that uses paired horizontal or vertical bars to compare quantities *see 4·2 Displaying Data*

Example:

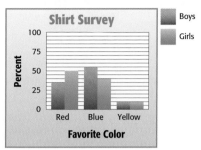

edge a line segment joining two planes of a polyhedron *see 6·2 Polygons and Polyhedrons*

English system units of measurement used in the United States that measure length in inches, feet, yards, and miles; capacity in cups, pints, quarts, and gallons; weight in ounces, pounds, and tons; and temperature in degrees Fahrenheit *see customary system*

equal angles angles that measure the same number of degrees *see congruent angles, 6·1 Naming and Classifying Angles and Triangles*

equally likely describes outcomes or events that have the same chance of occurring *see 4·5 Probability*

equally unlikely describes outcomes or events that have the same chance of not occurring *see 4·5 Probability*

equation a mathematical sentence stating that two expressions are equal *see 5·1 Writing Expressions and Equations, 5·8 Slope and Intercept*

Example: $3 \times (7 + 8) = 9 \times 5$

equiangular the property of a polygon in which all angles are congruent

equiangular triangle a triangle in which each angle is 60°
see 6·1 Naming and Classifying Angles and Triangles

Example:

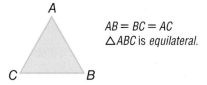

$m\angle A = m\angle B = m\angle C = 60°$
$\triangle ABC$ is equiangular.

equilateral the property of a polygon in which all sides are
congruent

equilateral triangle a triangle in which all sides are congruent

Example:

$AB = BC = AC$
$\triangle ABC$ is equilateral.

equivalent equal in value *see 2·1 Fractions and Equivalent
Fractions*

equivalent expressions expressions that always result in the
same number, or have the same mathematical meaning for all
replacement values of their variables *see 5·2 Simplifying
Expressions*

Examples: $\frac{9}{3} + 2 = 10 - 5$
$2x + 3x = 5x$

equivalent fractions fractions that represent the same
quotient but have different numerators and denominators
see 2·1 Fractions and Equivalent Fractions

Example: $\frac{5}{6} = \frac{15}{18}$

equivalent ratios ratios that are equal *see 5·5 Ratio and
Proportion*

Example: $\frac{5}{4} = \frac{10}{8}$; 5:4 = 10:8

estimate an approximation or rough calculation
see 2·6 Decimal Operations

even number any whole number that is a multiple of 2
{0, 2, 4, 6, 8, 10, 12 . . .}

event any happening to which probabilities can be assigned
see 4·5 Probability

expanded form a method of writing a number that highlights
the value of each digit

Example: $867 = (8 \times 100) + (6 \times 10) + (7 \times 1)$

expense an amount of money paid; cost

experimental probability the ratio of the total number of
times the favorable outcome occurs to the total number of
times the experiment is completed *see 4·5 Probability*

exponent a numeral that indicates how many times a number
or variable is used as a factor *see 1·3 Factors and Multiples,
3·1 Powers and Exponents, 3·3 Scientific Notation*

Example: In the equation $2^3 = 8$, the *exponent* is 3.

expression a mathematical combination of numbers, variables,
and operations *see 5·1 Writing Expressions and Equations,
5·2 Simplifying Expressions, 5·3 Evaluating Expressions and
Formulas*

Example: $6x + y^2$

face a two-dimensional side of a three-dimensional figure
see 6·2 Polygons and Polyhedrons, 6·6 Surface Area

factor a number or expression that is multiplied by another to
yield a product *see 1·3 Factors and Multiples, 2·4 Multiplication
and Division of Fractions, 3·1 Powers and Exponents*

Example: 3 and 11 are *factors* of 33.

factorial represented by the symbol !, the product of all the
whole numbers between 1 and a given positive whole number
see 4·5 Probability, 8·1 Scientific Calculator

Example: $5! = 1 \times 2 \times 3 \times 4 \times 5 = 120$

factor pair two unique numbers multiplied together to yield a product, such as $2 \times 3 = 6$ *see 1·3 Factors and Multiples*

fair describes a situation in which the theoretical probability of each outcome is equal

Fibonacci numbers *see Patterns page 63*

flip a transformation that produces the mirror image of a figure *see 6·3 Symmetry and Transformations*

Example:

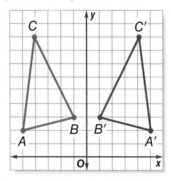

$\triangle A'B'C'$ is a *flip* of $\triangle ABC$.

formula an equation that shows the relationship between two or more quantities; a calculation performed by a spreadsheet *see Formulas pages 60–61, 5·3 Evaluating Expressions and Formulas, 8·3 Spreadsheets*

Example: $A = \pi r^2$ is the *formula* for calculating the area of a circle; A2 * B2 is a spreadsheet *formula*.

fraction a number representing part of a whole; a quotient in the form $\frac{a}{b}$ *see 2·1 Fractions and Equivalent Fractions*

function the assignment of exactly one output value to each input value

Example: You are driving at 50 miles per hour. There is a relationship between the amount of time you drive and the distance you will travel. You say that the distance is a *function* of the time.

geometric sequence *see Patterns page 63*

geometry the branch of mathematics that investigates the relations, properties, and measurements of solids, surfaces, lines, and angles *see Chapter 6 Geometry, 8·3 Geometry Tools*

gram a metric unit of mass *see 7·1 Systems of Measurement, 7·3 Area, Volume, and Capacity*

greatest common factor (GCF) the greatest number that is a factor of two or more numbers *see 1·3 Factors and Multiples, 2·1 Fractions and Equivalent Fractions*

Example: 30, 60, 75
The *greatest common factor* is 15.

harmonic sequence *see Patterns page 63*

height the perpendicular distance from a vertex to the opposite side of a figure *see 6·7 Volume*

heptagon a polygon with seven angles and seven sides *see 6·2 Polygons and Polyhedrons*

Example:

a *heptagon*

hexagon a polygon with six angles and six sides *see 6·2 Polygons and Polyhedrons*

Example:

a *hexagon*

hexagonal prism a prism that has two hexagonal bases and six rectangular sides *see 6·2 Polygons and Polyhedrons*

Example:

a *hexagonal prism*

hexahedron a polyhedron that has six faces *see 6·2 Polygons and Polyhedrons*

Example:

A cube is a *hexahedron.*

histogram a special kind of bar graph that displays the frequency of data that has been organized into equal intervals *see 4·2 Displaying Data*

horizontal parallel to or in the plane of the horizon

hypotenuse the side opposite the right angle in a right triangle *see 6·1 Naming and Classifying Angles and Triangles, 6·9 Pythagorean Theorem*

Example:

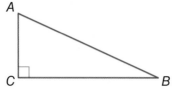

Side $\overline{AB}$ is the *hypotenuse* of this right triangle.

improper fraction a fraction in which the numerator is greater than the denominator *see 2·1 Fractions and Equivalent Fractions*

Examples: $\frac{21}{4}, \frac{4}{3}, \frac{2}{1}$

income the amount of money received for labor, services, or the sale of goods or property

independent event two events in which the outcome of one event is not affected by the outcome of another event *see 4·5 Probability*

inequality a statement that uses the symbols > (greater than), < (less than), ≥ (greater than or equal to), and ≤ (less than or equal to) to compare quantities *see 5·6 Inequalities*

Examples: $5 > 3; \frac{4}{5} < \frac{5}{4}; 2(5 - x) > 3 + 1$

infinite, nonrepeating decimal irrational numbers, such as π and $\sqrt{2}$, that are decimals with digits that continue indefinitely but do not repeat

inscribed figure a figure that is enclosed by another figure as shown below

Examples:

a triangle *inscribed* in a circle   a circle *inscribed* in a triangle

integers the set of all whole numbers and their additive inverses {. . ., −5, −4, −3, −2, −1, 0, 1, 2, 3, 4, 5, . . .}

intercept [1] the cutting of a line, curve, or surface by another line, curve, or surface; [2] the point at which a line or curve cuts across a coordinate axis *see 5·8 Slope and Intercept*

intersection the set of elements common to two or more sets
see Venn diagram, 1·3 Factors and Multiples

Example:

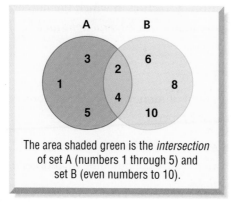

A   B

3   6
2
1   8
4
5   10

The area shaded green is the *intersection*
of set A (numbers 1 through 5) and
set B (even numbers to 10).

inverse operation the operation that reverses the effect of
another operation

Examples: Subtraction is the *inverse operation* of addition.
Division is the *inverse operation* of multiplication.

irrational numbers the set of all numbers that cannot be
expressed as finite or repeating decimals *see 2·5 The Real
Number System*

Example: $\sqrt{2}$ (1.414214 . . .) and π (3.141592 . . .) are *irrational
numbers.*

isometric drawing a two-dimensional representation of a
three-dimensional object in which parallel edges are drawn as
parallel lines

Example:

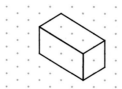

isosceles trapezoid a trapezoid in which the two nonparallel sides are of equal length

Example:

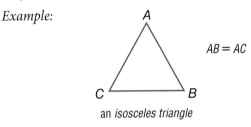

$AC = BD$

an *isosceles trapezoid*

isosceles triangle a triangle with at least two sides of equal length *see 6·1 Naming and Classifying Angles and Triangles*

Example:

$AB = AC$

an *isosceles triangle*

································ **L** ································

leaf the unit digit of an item of numerical data between 1 and 99 *see stem-and-leaf plot, 4·2 Displaying Data*

least common denominator (LCD) the least common multiple of the denominators of two or more fractions *see 2·3 Addition and Subtraction of Fractions*

Example: The *least common denominator* of $\frac{1}{3}$, $\frac{2}{4}$, and $\frac{3}{6}$ is 12.

least common multiple (LCM) the smallest nonzero whole number that is a multiple of two or more whole numbers *see 1·3 Factors and Multiples, 2·3 Addition and Subtraction of Fractions*

Example: The *least common multiple* of 3, 9, and 12 is 36.

legs of a triangle the sides adjacent to the right angle of a right triangle *see 7·9 Pythagorean Theorem*

Example:

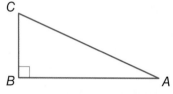

$\overline{AB}$ and $\overline{BC}$ are the *legs of* $\triangle ABC$.

length a measure of the distance of an object from end to end *see 7·2 Length and Distance*

likelihood the chance of a particular outcome occurring *see 4·5 Probability*

like terms terms that include the same variables raised to the same powers; *like terms* can be combined *see 5·2 Simplifying Expressions*

Example: $5x^2$ and $6x^2$ are *like terms*; $3xy$ and $3zy$ are not like terms.

line a connected set of points extending forever in both directions *see 6·1 Naming and Classifying Angles and Triangles*

linear equation an equation with two variables (x and y) that takes the general form $y = mx + b$, where m is the slope of the line and b is the y-intercept *see 5·4 Solving Linear Equations*

linear measure the measure of the distance between two points on a line

line graph a display of data that shows change over time
see 4·2 Displaying Data

Example:

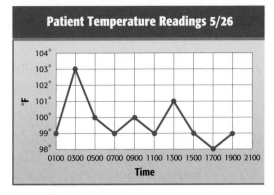

line of best fit on a scatter plot, a line drawn as near as possible to the various points so as to best represent the trend in data

Example:

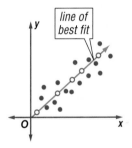

line of symmetry a line along which a figure can be folded so that the two resulting halves match *see 6·3 Symmetry and Transformations*

Example:

$\overline{ST}$ is a *line of symmetry.*

line plot a display of data that shows the frequency of data on a number line *see 4·2 Displaying Data*

line segment a section of a line between two points
see 6·1 Naming and Classifying Angles and Triangles

Example: A •————————• B

$\overline{AB}$ is a *line segment.*

liter a metric unit of capacity *see 7·3 Area, Volume, and Capacity*

lowest common multiple the smallest number that is a multiple of all the numbers in a given set; same as least common multiple *see 1·3 Factors and Multiples*

Example: The *lowest common multiple* of 6, 9, and 18 is 18.

Lucas numbers *see Patterns page 63*

magic square *see Patterns page 64*

maximum value the greatest value of a function or a set of numbers

mean the quotient obtained when the sum of the numbers in a set is divided by the number of addends *see average, 4·4 Statistics*

Example: The *mean* of 3, 4, 7, and 10 is
$(3 + 4 + 7 + 10) \div 4 = 6.$

measurement units standard measures, such as the meter, the liter, and the gram, or the foot, the quart, and the pound *see 7·1 Systems of Measurement*

median the middle number in an ordered set of numbers *see 4·4 Statistics*

Example: 1, 3, 9, 16, 22, 25, 27
16 is the *median.*

meter the metric unit of length

metric system a decimal system of weights and measurements based on the meter as its unit of length, the kilogram as its unit of mass, and the liter as its unit of capacity
see 7·1 Systems of Measurement

midpoint the point on a line segment that divides it into two equal segments

Example:

$AM = MB$

M is the *midpoint* of $\overline{AB}$.

minimum value the least value of a function or a set of numbers

mixed number a number composed of a whole number and a fraction *see 2·1 Fractions and Equivalent Fractions*

Example: $5\frac{1}{4}$ is a *mixed number.*

mode the number or element that occurs most frequently in a set of data *see 4·4 Statistics*

Example: 1, 1, 2, 2, 3, 5, 5, 6, 6, 6, 8
6 is the *mode.*

monomial an algebraic expression consisting of a single term

Example: $5x^3y$, xy, and $2y$ are three *monomials.*

multiple the product of a given number and an integer
see 1·3 Factors and Multiples

Examples: 8 is a *multiple* of 4.
3.6 is a *multiple* of 1.2.

multiplication one of the four basic arithmetical operations, involving the repeated addition of numbers

multiplication growth number a number that when used to multiply a given number a given number of times results in a given goal number

Example: Grow 10 into 40 in two steps by multiplying
$(10 \times 2 \times 2 = 40)$.
2 is the *multiplication growth number.*

Multiplication Property of Equality the mathematical rule that states that if each side of an equation is multiplied by the same number, the two sides remain equal *see 5·4 Solving Linear Equations*

Example: If $a = b$, then $a \times c = b \times c$.

multiplicative inverse two numbers are multiplicative inverses if their product is 1 *see 2·4 Multiplication and Division of Fractions*

Example: $10 \times \dfrac{1}{10} = 1$

$\dfrac{1}{10}$ is the *multiplicative inverse* of 10.

···································· **N** ····································

natural variability the difference in results in a small number of experimental trials from the theoretical probabilities

negative integers the set of all integers that are less than zero $\{-1, -2, -3, -4, -5, \ldots\}$ *see 1·4 Integer Operations*

negative numbers the set of all real numbers that are less than zero $\{-1, -1.36, -\sqrt{2}, -\pi\}$ *see 1·4 Integer Operations*

net a two-dimensional plan that can be folded to make a three-dimensional model of a solid *see 6·6 Surface Area*

Example:

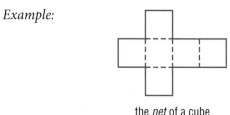

the *net* of a cube

nonagon a polygon with nine angles and nine sides *see 6·2 Polygons and Polyhedrons*

Example:

a *nonagon*

noncollinear points not lying on the same line

noncoplanar points or lines not lying on the same plane

number line a line showing numbers at regular intervals on which any real number can be indicated *see 5·6 Inequalities*

Example:

a *number line*

number symbols the symbols used in counting and measuring

Examples: $1, -\frac{1}{4}, 5, \sqrt{2}, -\pi$

numerator the top number in a fraction representing the number of equal parts being considered *see 2·1 Fractions and Equivalent Fractions*

Example: In the fraction $\frac{a}{b}$, a is the *numerator*.

.... **O** ....

obtuse angle any angle that measures greater than 90° but less than 180° *see 6·1 Naming and Classifying Angles and Triangles*

Example:

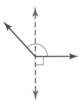

an *obtuse angle*

obtuse triangle a triangle that has one obtuse angle *see 6·1 Naming and Classifying Angles and Triangles*

Example:

△ABC is an *obtuse triangle*.

octagon a polygon with eight angles and eight sides
see 6·2 Polygons and Polyhedrons

Example:

an *octagon*

octagonal prism a prism that has two octagonal bases and
eight rectangular faces *see 6·2 Polygons and Polyhedrons*

Example:

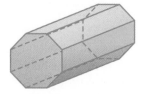

an *octagonal prism*

odd numbers the set of all integers that are not multiples of 2

odds against the ratio of the number of unfavorable outcomes
to the number of favorable outcomes

odds for the ratio of the number of favorable outcomes to the
number of unfavorable outcomes

one-dimensional having only one measurable quality
see Chapter 6 Geometry

Example: A line and a curve are *one-dimensional.*

operations arithmetical actions performed on numbers,
matrices, or vectors

opposite angle in a triangle, a side and an angle are said to be *opposite* if the side is not used to form the angle

Example:

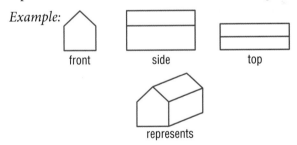

In △*ABC*, ∠*A* is opposite of $\overline{BC}$.

ordered pair two numbers that tell the *x*-coordinate and *y*-coordinate of a point *see 5·7 Graphing on the Coordinate Plane*

Example: The coordinates (3, 4) are an *ordered pair*. The *x*-coordinate is 3, and the *y*-coordinate is 4.

order of operations to simplify an expression, follow this four-step process: 1) do all operations within parentheses; 2) simplify all numbers with exponents; 3) multiply and divide in order from left to right; 4) add and subtract in order from left to right *see 1·2 Order of Operations*

origin the point (0, 0) on a coordinate graph where the *x*-axis and the *y*-axis intersect

orthogonal drawing always shows three views of an object—top, side, and front; the views are drawn straight-on

Example:

front   side   top

represents

outcome a possible result in a probability experiment *see 4·5 Probability*

outcome grid a visual model for analyzing and representing theoretical probabilities that shows all the possible outcomes of two independent events *see 4·5 Probability*

Example:

A grid used to find the sample space for rolling a pair of dice. The outcomes are written as ordered pairs.

	1	2	3	4	5	6
1	(1, 1)	(2, 1)	(3, 1)	(4, 1)	(5, 1)	(6, 1)
2	(1, 2)	(2, 2)	(3, 2)	(4, 2)	(5, 2)	(6, 2)
3	(1, 3)	(2, 3)	(3, 3)	(4, 3)	(5, 3)	(6, 3)
4	(1, 4)	(2, 4)	(3, 4)	(4, 4)	(5, 4)	(6, 4)
5	(1, 5)	(2, 5)	(3, 5)	(4, 5)	(5, 5)	(6, 5)
6	(1, 6)	(2, 6)	(3, 6)	(4, 6)	(5, 6)	(6, 6)

There are 36 possible outcomes.

outlier data that are more than 1.5 times the interquartile range from the upper or lower quartiles *see 4·4 Statistics*

·· **P** ····································

parallel straight lines or planes that remain a constant distance from each other and never intersect, represented by the symbol ∥

Example:

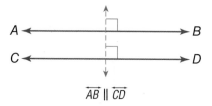

$$\overline{AB} \parallel \overline{CD}$$

parallelogram a quadrilateral with two pairs of parallel sides *see 6·2 Polygons and Polyhedrons*

Example:

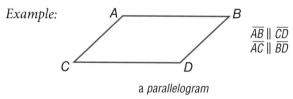

$$\overline{AB} \parallel \overline{CD}$$
$$\overline{AC} \parallel \overline{BD}$$

a *parallelogram*

parentheses the enclosing symbols ( ), which indicate that the terms within are a unit

Example: $(2 + 4) \div 2 = 3$

Pascal's Triangle *see Patterns page 64*

pattern a regular, repeating design or sequence of shapes or numbers *see Patterns pages 63–65*

PEMDAS an acronym for the order of operations: 1) do all operations within **p**arentheses; 2) simplify all numbers with **e**xponents; 3) **m**ultiply and **d**ivide in order from left to right; 4) **a**dd and **s**ubtract in order from left to right *see 1·2 Order of Operations*

pentagon a polygon with five angles and five sides *see 6·2 Polygons and Polyhedrons*

Example:

a *pentagon*

percent a number expressed in relation to 100, represented by the symbol % *see 2·7 Meaning of Percents, 4·2 Analyzing Data*

Example: 76 out of 100 students use computers.
76 *percent* or 76% of students use computers.

percent grade the ratio of the rise to the run of a hill, ramp, or incline expressed as a percent

Example:

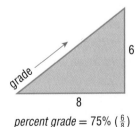

percent grade = 75% ($\frac{6}{8}$)

percent proportion compares part of a quantity to the whole quantity using a percent

$$\frac{part}{whole} = \frac{percent}{100}$$

perfect cube a number that is the cube of an integer

Example: 27 is a *perfect cube* since $27 = 3^3$.

perfect number an integer that is equal to the sum of all its positive whole number divisors, excluding the number itself

Example: $1 \times 2 \times 3 = 6$ and $1 + 2 + 3 = 6$
  6 is a *perfect number.*

perfect square a number that is the square of an integer
see *3·2 Square Roots*

Example: 25 is a *perfect square* since $25 = 5^2$.

perimeter the distance around the outside of a closed figure
see *Formulas page 60, 6·4 Perimeter*

Example:

$AB + BC + CD + DA = perimeter$

permutation a possible arrangement of a group of objects; the number of possible arrangements of *n* objects is expressed by the term *n*! *see factorial, 4·5 Probability*

perpendicular two lines or planes that intersect to form a right angle

Example:

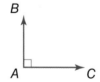

$\overline{AB}$ and $\overline{AC}$ are *perpendicular.*

pi the ratio of a circle's circumference to its diameter; *pi* is shown by the symbol π, and is approximately equal to 3.14
see *7·8 Circles*

picture graph a display of data that uses pictures or symbols to represent numbers *see 4·2 Displaying Data*

place value the value given to a place a digit occupies in a numeral *see 2·1 Naming and Ordering Decimals*

place-value system a number system in which values are given to the places digits occupy in the numeral; in the decimal system, the value of each place is 10 times the value of the place to its right *see 2·1 Naming and Ordering Decimals*

point one of four undefined terms in geometry used to define all other terms; a *point* has no size *see 5·7 Graphing on the Coordinate Plane*

polygon a simple, close plane figure, having three or more line segments as sides *see 6·2 Polygons and Polyhedrons*

Examples:

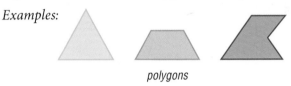

polygons

polyhedron a solid geometrical figure that has four or more plane faces *see 6·2 Polygons and Polyhedrons*

Examples:

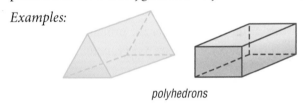

polyhedrons

population the universal set from which a sample of statistical data is selected *see 4·1 Collecting Data*

positive integers the set of all integers that are greater than zero {1, 2, 3, 4, 5, . . .} *see 1·4 Integer Operations*

positive numbers the set of all real numbers that are greater than zero {1, 1.36, $\sqrt{2}$, π}

power represented by the exponent n, to which a number is raised as a factor n times *see 3·1 Powers and Exponents*

Example: 7 raised to the fourth *power.*
$$7^4 = 7 \times 7 \times 7 \times 7 = 2{,}401$$

predict to anticipate a trend by studying statistical data

prime factorization the expression of a composite number as a product of its prime factors *see 1·3 Factors and Multiples*

Examples: $504 = 2^3 \times 3^2 \times 7$
$30 = 2 \times 3 \times 5$

prime number a whole number greater than 1 whose only factors are 1 and itself *see 1·3 Factors and Multiples*

Examples: 2, 3, 5, 7, 11

prism a solid figure that has two parallel, congruent polygonal faces (called *bases*) *see 6·2 Polygons and Polyhedrons*

Examples:

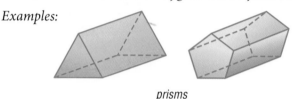

prisms

probability the study of likelihood or chance that describes the possibility of an event occurring *see 4·5 Probability*

probability line a line used to order the probability of events from least likely to most likely *see 4·5 Probability*

probability of events the likelihood or chance that events will occur *see 4·5 Probability*

product the result obtained by multiplying two numbers or variables *see 2·4 Multiplication and Division of Factors*

profit the gain from a business; what is left when the cost of goods and of carrying on the business is subtracted from the amount of money taken in

project to extend a numerical model, to either greater or lesser values, in order to predict likely quantities in an unknown situation

proportion a statement that two ratios are equal *see 5·5 Ratio and Proportion*

pyramid a solid geometrical figure that has a polygonal base and triangular faces that meet at a common vertex *see 6·2 Polygons and Polyhedrons*

Examples:

pyramids

Pythagorean Theorem a mathematical idea stating that the sum of the squared lengths of the two legs of a right triangle is equal to the squared length of the hypotenuse *see 6·9 Pythagorean Theorem*

Example:

For a right triangle, $a^2 + b^2 = c^2$.

Pythagorean triple a set of three positive integers a, b, and c, such that $a^2 + b^2 = c^2$ *see 6·9 Pythagorean Theorem*

Example: The Pythagorean triple {3, 4, 5}
$$3^2 + 4^2 = 5^2$$
$$9 + 16 = 25$$

·················· **Q** ··················

quadrant [1] one quarter of the circumference of a circle; [2] on a coordinate graph, one of the four regions created by the intersection of the x-axis and the y-axis *see 5·7 Graphing on the Coordinate Plane*

quadrilateral a polygon that has four sides *see 6·2 Polygons and Polyhedrons*

Examples:

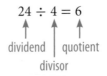

quadrilaterals

qualitative graphs a graph with words that describes such things as a general trend of profits, income, and expenses over time; it has no specific numbers

quantitative graphs a graph that, in contrast to a qualitative graph, has specific numbers

quotient the result obtained from dividing one number or variable (the divisor) into another number or variable (the dividend) *see 5·1 Writing Expressions and Equations*

Example:

$$24 \div 4 = 6$$

dividend   quotient

divisor

radical the indicated root of a quantity

Examples: $\sqrt{3}, \sqrt[4]{14}, \sqrt[12]{23}$

radical sign the root symbol $\sqrt{}$

radius a line segment from the center of a circle to any point on its perimeter *see 6·8 Circles*

random sample a population sample chosen so that each member has the same probability of being selected *see 4·1 Collecting Data*

range in statistics, the difference between the largest and smallest values in a sample *see 4·4 Statistics*

rank to order the data from a statistical sample on the basis of some criterion—for example, in ascending or descending numerical order

ranking the position on a list of data from a statistical sample based on some criterion

rate [1] fixed ratio between two things; [2] a comparison of two different kinds of units, for example, miles per hour or dollars per hour *see 5·5 Ratio and Proportion*

ratio a comparison of two numbers *see 2·7 Meaning of Percent, 5·5 Ratio and Proportion*

Example: The *ratio* of consonants to vowels in the alphabet is 21:5.

rational numbers the set of numbers that can be written in the form $\frac{a}{b}$, where a and b are integers and b does not equal zero *see 2·9 Fraction, Decimal, and Percent Relationships*

Examples: $1 = \frac{1}{1}, \frac{2}{9}, 3\frac{2}{7} = \frac{23}{7}, -0.333 = -\frac{1}{3}$

ray the part of a straight line that extends infinitely in one direction from a fixed point *see 6·1 Naming and Classifying Angles and Triangles*

Example:

a *ray*

real numbers the set consisting of zero, all positive numbers, and all negative numbers; *real numbers* include all rational and irrational numbers

real-world data information processed by people in everyday situations

reciprocal one of a pair of numbers that have a product of 1 *see 2·4 Multiplication and Division of Fractions*

Examples: The *reciprocal* of 2 is $\frac{1}{2}$; of $\frac{3}{4}$ is $\frac{4}{3}$; of x is $\frac{1}{x}$.

rectangle a parallelogram with four right angles
see 6·2 Polygons and Polyhedrons

Example:

a *rectangle*

rectangular prism a prism that has rectangular bases and four rectangular faces *see 6·2 Polygons and Polyhedrons*

reflection a transformation that produces the mirror image of a figure *see 6·3 Symmetry and Transformations*

Example:

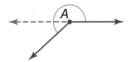

the *reflection* of a trapezoid

reflex angle any angle with a measure that is greater than 180° but less than 360° *see 6·1 Naming and Classifying Angles and Triangles*

Example:

∠A is a *reflex angle.*

regular polygon a polygon in which all sides are equal and all angles are congruent *see 6·2 Polygons and Polyhedrons*

a *regular polygon*

relationship a connection between two or more objects, numbers, or sets; a mathematical *relationship* can be expressed in words or with numbers and letters

repeating decimal a decimal in which a digit or a set of digits repeat infinitely *see 2·9 Fraction, Decimal, and Percent Relationships*

Example: 0.121212 . . . is a *repeating decimal.*

rhombus a parallelogram with all sides of equal length *see 6·2 Polygons and Polyhedrons*

Example:

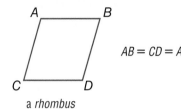

$AB = CD = AC = BD$

a *rhombus*

right angle an angle that measures 90° *see 6·1 Naming and Classifying Angles and Triangles, 6·9 Pythagorean Theorem*

Example:

∠A is a *right angle.*

right triangle a triangle with one right angle *see 6·1 Naming and Classifying Angles and Triangles*

Example:

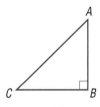

△ABC is a *right triangle.*

rise the vertical distance between two points *see 5·8 Slope and Intercept*

Roman numerals the numeral system consisting of the symbols I (1), V (5), X (10), L (50), C (100), D (500), and M (1,000); when a Roman symbol is preceded by a symbol of equal or greater value, the values of a symbol are added (XVI = 16); when a symbol is preceded by a symbol of lesser value, the values are subtracted (IV = 4)

root [1] the inverse of an exponent; [2] the radical sign $\sqrt{}$ indicates square root *see 3·2 Square Roots, 8·1 Scientific Calculator*

rotation a transformation in which a figure is turned a certain number of degrees around a fixed point or line *see 6·3 Symmetry and Transformations*

Example:

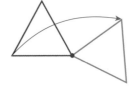

the *turning* of a triangle

round to approximate the value of a number to a given decimal place *see 2·6 Decimal Operations*

Examples: 2.56 rounded to the nearest tenth is 2.6;
2.54 rounded to the nearest tenth is 2.5;
365 rounded to the nearest hundred is 400.

row a horizontal list of numbers or terms *see 8·3 Spreadsheets*

rule a statement that describes a relationship between numbers or objects

run the horizontal distance between two points *see 5·8 Slope and Intercept*

sample a finite subset of a population, used for statistical analysis *see 4·1 Collecting Data*

sample space the set of all possible outcomes of a probability experiment *see 4·5 Combinations and Permutations*

scale the ratio between the actual size of an object and a proportional representation *see 7·5 Size and Scale*

scale drawing a proportionally correct drawing of an object or area at actual, enlarged, or reduced size *see 7·5 Size and Scale*

scale factor the factor by which all the components of an object are multiplied in order to create a proportional enlargement or reduction *see 7·5 Size and Scale*

scalene triangle a triangle with no sides of equal length *see 6·1 Naming and Classifying Angles and Triangles*

Example:

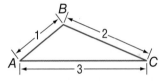

△*ABC* is a *scalene triangle.*

scale size the proportional size of an enlarged or reduced representation of an object or area *see 7·5 Size and Scale*

scatter plot (or scatter diagram) a display of data in which the points corresponding to two related factors are graphed and observed for correlation *see 4·3 Analyzing Data*

Example:

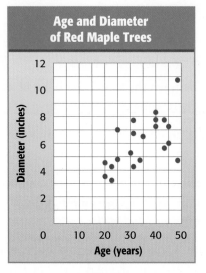

scatter plot

scientific notation a method of writing a number using exponents and powers of ten; a number in scientific notation is written as a number between 1 and 10 multiplied by a power of ten *see 3·3 Scientific Notation*

Examples: $9{,}572 = 9.572 \times 10^{3}$ and $0.00042 = 4.2 \times 10^{-4}$

segment two points and all the points on the line between them *see 6·2 Angles and Triangles*

sequence *see Patterns page 64*

series *see Patterns page 64*

set a collection of distinct elements or items

side a line segment that forms an angle or joins the vertices of a polygon *see 6·4 Perimeter*

sighting measuring a length or angle of an inaccessible object by lining up a measuring tool with one's line of vision

signed number a number preceded by a positive or negative sign *see 1·4 Integer Operations*

significant digit the number of digits in a value that indicate its precision and accuracy

Example: 297,624 rounded to three significant digits is 298,000; 2.97624 rounded to three significant digits is 2.98.

similar figures have the same shape but are not necessarily the same size *see 7·5 Size and Scale*

Example:

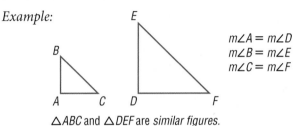

$m\angle A = m\angle D$
$m\angle B = m\angle E$
$m\angle C = m\angle F$

△*ABC* and △*DEF* are *similar figures.*

simple event an outcome or collection of outcomes *see 4·5 Probability*

simulation a mathematical experiment that approximates real-world processes

slide to move a shape to another position without rotating or reflecting it; also referred to as a translation *see 6·3 Symmetry and Transformations*

Example:

the *slide* of a trapezoid

slope [1] a way of describing the steepness of a line, ramp, hill, and so on; [2] the ratio of the rise to the run *see 5·8 Slope and Intercept*

slope angle the angle that a line forms with the *x*-axis or other horizontal

slope ratio the slope of a line as a ratio of the rise to the run

solid a three-dimensional figure

solution the answer to a mathematical problem; in algebra, a *solution* usually consists of a value or set of values for a variable

speed the rate at which an object moves

speed-time graph a graph used to chart how the speed of an object changes over time

sphere a perfectly round geometric solid, consisting of a set of points equidistant from a center point

Example:

a *sphere*

spinner a device for determining outcomes in a probability experiment *see 4·5 Probability*

Example:

a *spinner*

spiral *see Patterns page 65*

spreadsheet a computer tool where information is arranged into cells within a grid and calculations are performed within the cells; when one cell is changed, all other cells that depend on it automatically change *see 8·3 Spreadsheets*

square [1] a rectangle with congruent sides; [2] the product of two equal terms *see 3·1 Powers and Exponents, 6·2 Polygons and Polyhedrons, 8·1 Scientific Calculator*

Examples: [1]

$AB = CD = AC = BD$

a *square*

[2] $4^2 = 4 \times 4 = 16$

square centimeter a unit used to measure the area of a surface; the area of a square measuring one centimeter on each side *see 7·3 Area, Volume, and Capacity*

square foot a unit used to measure the area of a surface; the area of a square measuring one foot on each side *see 7·3 Area, Volume, and Capacity*

square inch a unit used to measure the area of a surface; the area of a square measuring one inch on each side *see 7·3 Area, Volume, and Capacity*

square meter a unit used to measure the area of a surface; the area of a square measuring one meter on each side *see 7·3 Area, Volume, and Capacity*

square number *see Patterns page 65*

square pyramid a pyramid with a square base

square root a number that when multiplied by itself equals a given number *see 3·2 Square Roots, 8·1 Scientific Calculator*

Example: 3 is the *square root* of 9.
$$\sqrt{9} = 3$$

square root symbol the mathematical symbol $\sqrt{}$; indicates that the square root of a given number is to be calculated *see 3·2 Square Roots*

standard measurement commonly used measurements, such as the meter used to measure length, the kilogram used to measure mass, and the second used to measure time *see Chapter 7 Measurement*

statistics the branch of mathematics that investigates the collection and analysis of data *see 4·3 Statistics*

steepness a way of describing the amount of incline (or slope) of a ramp, hill, line, and so on

stem the tens digit of an item of numerical data between 1 and 99 *see stem-and-leaf plot, 4·2 Displaying Data*

stem-and-leaf plot a method of displaying numerical data between 1 and 99 by separating each number into its tens digit (stem) and its unit digit (leaf) and then arranging the data in ascending order of the tens digits *see 4·2 Displaying Data*

Example:

Average Points per Game

Stem	Leaf
0	6
1	1 8 2 2 5
2	6 1
3	7
4	3
5	8

2 | 6 = *26 points*

a *stem-and-leaf plot* for the data set
11, 26, 18, 12, 12, 15, 43, 37, 58, 6, and 21

straight angle an angle that measures 180°; a straight line

subtraction one of the four basic arithmetical operations, taking one number or quantity away from another

Subtraction Property of Equality the mathematical rule that states that if the same number is subtracted from each side of the equation, then the two sides remain equal *see 6·4 Solving Linear Equations*

Example: If $a = b$, then $a - c = b - c$.

sum the result of adding two numbers or quantities *see 5·1 Writing Expressions and Equations*

Example: $6 + 4 = 10$

$\qquad$ 10 is the *sum* of the two addends, 6 and 4.

supplementary angles two angles that have measures whose sum is 180° *see 6·1 Naming and Classifying Angles and Triangles*

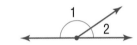

∠1 and ∠2 are *supplementary angles.*

surface area the sum of the areas of all the faces of a geometric solid, measured in square units *see 6·6 Surface Area*

Example:

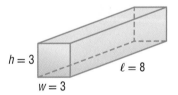

The *surface area* of this rectangular prism is
$2(3 \times 3) + 4(3 \times 8) = 114$ square units.

survey a method of collecting statistical data in which people are asked to answer questions *see 4·1 Collecting Data*

symmetry *see line of symmetry, 6·3 Symmetry and Transformations*

Example:

This hexagon has *symmetry* around the dotted line.

table a collection of data arranged so that information can be easily seen *see 4·2 Displaying Data*

tally marks marks made for certain numbers of objects in keeping account *see 4·1 Collecting Data*

Example: ⦀⦀ ⦀⦀⦀ = 8

term product of numbers and variables *see Chapter 5 Algebra*

Example: x, ax^2, $2x^4y^2$, and $-4ab$

terminating decimal a decimal with a finite number of digits *see 2·9 Fraction, Decimal, and Percent Relationships*

tessellation *see Patterns page 65*

tetrahedron a geometrical solid that has four triangular faces *see 6·2 Polygons and Polyhedrons*

Example:

a *tetrahedron*

theoretical probability the ratio of the number of favorable outcomes to the total number of possible outcomes *see 4·5 Probability*

three-dimensional having three measurable qualities: length, height, and width

tiling completely covering a plane with geometric shapes *see tessellations*

time in mathematics, the element of duration, usually represented by the variable t

total distance the amount of space between a starting point and an endpoint, represented by d in the equation $d = s$ (speed) $\times t$ (time)

total distance graph a coordinate graph that shows cumulative distance traveled as a function of time

total time the duration of an event, represented by t in the equation $t = \dfrac{d \ (\text{distance})}{s \ (\text{speed})}$

transformation a mathematical process that changes the shape or position of a geometric figure *see 6·3 Symmetry and Transformations*

translation a transformation in which a geometric figure is slid to another position without rotation or reflection *see 6·3 Symmetry and Transformations*

trapezoid a quadrilateral with only one pair of parallel sides *see 6·2 Polygons and Polyhedrons*

Example:

a *trapezoid*

tree diagram a connected, branching graph used to diagram probabilities or factors *see prime factorization, 4·5 Probability*

Example:

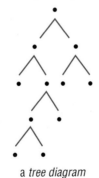

a *tree diagram*

trend a consistent change over time in the statistical data representing a particular population

triangle a polygon with three angles and three sides *see 6·1 Naming and Classifying Angles and Triangles*

triangular numbers *see Patterns page 65*

triangular prism a prism with two triangular bases and three rectangular sides *see prism*

turn to move a geometric figure by rotating it around a point *see rotation, 6·3 Symmetry and Transformations*

Example:

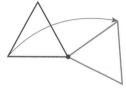

the *turning* of a triangle

two-dimensional having two measurable qualities: length and width

· (U) ·

unbiased sample a sample representative of the entire population *see 4·1 Collecting Data*

unequal probabilities different likelihoods of occurrence; two events have *unequal probabilities* if one is more likely to occur than the other

unfair where the probability of each outcome is not equal

union a set that is formed by combining the members of two or more sets, as represented by the symbol ∪; the *union* contains all members previously contained in both sets *see Venn diagram, 1·3 Factors and Multiples*

Example:

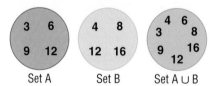

Set A   Set B   Set A ∪ B

The orange circle shows the *union* of sets A and B.

unit price the price of a single item or amount

unit rate the rate in lowest terms

Example: 120 miles in two hours is equivalent to a *unit rate* of 60 miles per hour.

variable a letter or other symbol that represents a number or set of numbers in an expression or an equation *see 5·1 Writing Expressions and Equations*

Example: In the equation $x + 2 = 7$, the variable is x.

Venn diagram a pictorial means of representing the relationships between sets *see 1·3 Factors and Multiples*

Example:

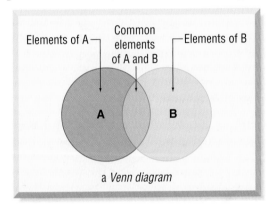

a *Venn diagram*

vertex (pl. *vertices*) the common point of two rays of an angle, two sides of a polygon, or three or more faces of a polyhedron

Examples:

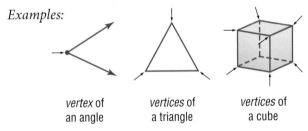

vertex of an angle   *vertices* of a triangle   *vertices* of a cube

vertex of tessellation the point where three or more tessellating figures come together

Example:

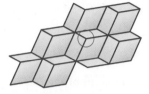

vertex of tessellation
(in the circle)

vertical a line that is perpendicular to a horizontal base line

Example:

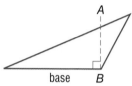

base   B

$\overline{AB}$ is *vertical* to the base
of this triangle.

vertical angles opposite angles formed by the intersection of two lines *see 6·1 Naming and Classifying Angles and Triangles*

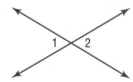

∠1 and ∠2 are *vertical angles.*

volume the space occupied by a solid, measured in cubic units *see Formulas page 60, 6·7 Volume*

Example:

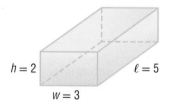

$h = 2$ $\ell = 5$

$w = 3$

The *volume* of this rectangular prism is 30 cubic units.
$2 \times 3 \times 5 = 30$

whole numbers the set of all counting numbers plus zero
{0, 1, 2, 3, 4, 5}

width a measure of the distance of an object from side to side

. . . . **X** . . . .

***x*-axis** the horizontal reference line in the coordinate graph
see 5·7 Graphing on the Coordinate Plane

***x*-intercept** the point at which a line or curve crosses the *x*-axis

. . . . **Y** . . . .

***y*-axis** the vertical reference line in the coordinate graph
see 5·7 Graphing on a Coordinate Plane

***y*-intercept** the point at which a line or curve crosses the
y-axis *see 5·8 Slope and Intercept*

. . . . **Z** . . . .

zero-pair one positive cube and one negative cube used to
model signed-number arithmetic

Formulas

Area (see 6·5)

circle	$A = \pi r^2$ (pi × square of the radius)
parallelogram	$A = bh$ (base × height)
rectangle	$A = \ell w$ (length × width)
square	$A = s^2$ (side squared)
trapezoid	$A = \frac{1}{2}h(b_1 + b_2)$
	($\frac{1}{2}$ × height × sum of the bases)
triangle	$A = \frac{1}{2}bh$
	($\frac{1}{2}$ × base × height)

Volume (see 6·7)

cone	$V = \frac{1}{3}\pi r^2 h$
	($\frac{1}{3}$ × pi × square of the radius × height)
cylinder	$V = \pi r^2 h$
	(pi × square of the radius × height)
prism	$V = Bh$ (area of the base × height)
pyramid	$V = \frac{1}{3}Bh$
	($\frac{1}{3}$ × area of the base × height)
rectangular prism	$V = \ell wh$ (length × width × height)
sphere	$V = \frac{4}{3}\pi r^3$
	($\frac{4}{3}$ × pi × cube of the radius)

Perimeter (see 6·4)

parallelogram	$P = 2a + 2b$
	(2 × side a + 2 × side b)
rectangle	$P = 2\ell + 2w$ (twice length + twice width)
square	$P = 4s$
	(4 × side)
triangle	$P = a + b + c$ (side a + side b + side c)

Formulas

Circumference (see 6·8)

circle $C = \pi d$ (pi × diameter)

or

$C = 2\pi r$

(2 × pi × radius)

Probability (see 4·5)

The *Experimental Probability* of an event is equal to the total number of times a favorable outcome occurred, divided by the total number of times the experiment was done.

$$Experimental\ Probability = \frac{favorable\ outcomes\ that\ occurred}{total\ number\ of\ experiments}$$

The *Theoretical Probability* of an event is equal to the number of favorable outcomes, divided by the total number of possible outcomes.

$$Theoretical\ Probability = \frac{favorable\ outcomes}{possible\ outcome}$$

Other

Distance $d = rt$ (rate × time)

Interest $I = prt$ (principle × rate × time)

PIE   Profit = Income − Expenses

Temperature $F = \frac{9}{5}C + 32$

($\frac{9}{5}$ × Temperature in °C + 32)

$C = \frac{5}{9}(F - 32)$

($\frac{5}{9}$ × (Temperature in °F − 32))

Symbols

{ }	set	$\overline{AB}$	segment AB
∅	the empty set	$\overrightarrow{AB}$	ray AB
⊆	is a subset of	$\overleftrightarrow{AB}$	line AB
∪	union	$\triangle ABC$	triangle ABC
∩	intersection	$\angle ABC$	angle ABC
>	is greater than	$m\angle ABC$	measure of angle ABC
<	is less than		
≥	is greater than or equal to	$\overline{AB}$ or $m\overline{AB}$	length of segment AB
≤	is less than or equal to	$\overgroup{AB}$	arc AB
=	is equal to	!	factorial
≠	is not equal to	$_nP_r$	permutations of n things taken r at a time
°	degree		
%	percent	$_nC_r$	combinations of n things taken r at a time
$f(n)$	function, f of n		
$a{:}b$	ratio of a to b, $\frac{a}{b}$	$\sqrt{}$	square root
$\lvert a\rvert$	absolute value of a	$\sqrt[3]{}$	cube root
$P(E)$	probability of an event E	$'$	foot
π	pi	$''$	inch
⊥	is perpendicular to	÷	divide
∥	is parallel to	/	divide
≅	is congruent to	*	multiply
∼	is similar to	×	multiply
≈	is approximately equal to	·	multiply
∠	angle	+	add
∟	right angle	−	subtract
△	triangle		

Patterns

arithmetic sequence a sequence of numbers or terms that have a common difference between any one term and the next in the sequence; in the following sequence, the common difference is seven, so $8 - 1 = 7$; $15 - 8 = 7$; $22 - 15 = 7$, and so forth

Example: 1, 8, 15, 22, 29, 36, 43, . . .

Fibonacci numbers a sequence in which each number is the sum of its two predecessors; can be expressed as $x_n = x_{n-2} + x_{n-1}$; the sequence begins: 1, 1, 2, 3, 5, 8, 13, 21, 34, 55, . . .

Example:

1,	1,	2,	3,	5,	8,	13,	21,	34,	55,	...
$1 + 1 = 2$										
	$1 + 2 = 3$									
		$2 + 3 = 5$								
			$3 + 5 = 8$							

geometric sequence a sequence of terms in which each term is a constant multiple, called the *common ratio,* of the one preceding it; for instance, in nature, the reproduction of many single-celled organisms is represented by a progression of cells splitting in two in a growth progression of 1, 2, 4, 8, 16, 32, . . ., which is a geometric sequence in which the common ratio is 2

harmonic sequence a progression $a_1, a_2, a_3, \ldots$ for which the reciprocals of the terms, $\frac{1}{a_1}, \frac{1}{a_2}, \frac{1}{a_3}, \ldots$ form an arithmetic sequence

Lucas numbers a sequence in which each number is the sum of its two predecessors; can be expressed as $x_n = x_{n-2} + x_{n-1}$; the sequence begins: 2, 1, 3, 4, 7, 11, 18, 29, 47, . . .

magic square a square array of different integers in which the sum of the rows, columns, and diagonals are the same

Example:

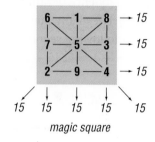

magic square

Pascal's triangle a triangular arrangement of numbers in which each number is the sum of the two numbers above it in the preceding row

Example:

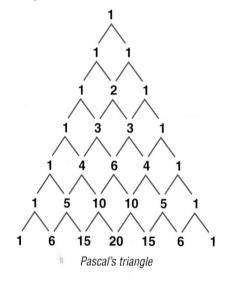

Pascal's triangle

sequence a set of elements, especially numbers, arranged in order according to some rule

series the sum of the terms of a sequence

spiral a plane curve traced by a point moving around a fixed point while continuously increasing or decreasing its distance from it

Example:

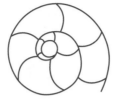

The shape of a chambered nautilus shell is a *spiral.*

square numbers a sequence of numbers that can be shown by dots arranged in the shape of a square; can be expressed as x^2; the sequence begins 1, 4, 9, 16, 25, 36, 49, . . .

Example:

| 1 | 4 | 9 | 16 | 25 | 36 |

square numbers

tessellation a tiling pattern made of repeating polygons that fills a plane completely, leaving no gaps

Examples:

tessellations

triangular numbers a sequence of numbers that can be shown by dots arranged in the shape of a triangle; any number in the sequence can be expressed as $x_n = x_{n-1} + n$; the sequence begins 1, 3, 6, 10, 15, 21, . . .

Example:

| 1 | 3 | 6 | 10 |

triangular numbers

estimate

height

probab

statistics

time

co

Hot Topics

Numbers and Computation

What do you know?

You can use the problems and the list of words that follow to see what you already know about this chapter. The answers to the problems are in **HotSolutions** at the back of the book, and the definitions of the words are in **HotWords** at the front of the book. You can find out more about a particular problem or word by referring to the topic number (*for example,* Lesson 1·2).

Problem Set

Solve. (Lesson 1·1)

1. 298×0 **2.** $(6 \times 3) \times 1$ **3.** $4,089 + 0$ **4.** 0×1

Solve using mental math. (Lesson 1·1)

5. $5 \times (31 + 69)$ **6.** $25 \times 14 \times 4$

Use parentheses to make each expression true. (Lesson 1·2)

7. $4 + 7 \times 4 = 44$ **8.** $20 + 16 \div 4 + 5 = 29$

Is it a prime number? Write *yes* or *no*. (Lesson 1·3)

9. 57   **10.** 102   **11.** 151   **12.** 203

Write the prime factorization for each number. (Lesson 1·3)

13. 35   **14.** 115   **15.** 220

Find the GCF for each pair of numbers. (Lesson 1·3)

16. 12 and 30   **17.** 15 and 60   **18.** 18 and 150

Find the LCM for each pair of numbers. (Lesson 1·3)

19. 5 and 15   **20.** 25 and 8   **21.** 18 and 40

22. A mystery number is a common multiple of 2, 3, and 12. It is a factor of 108. What is the number? (Lesson 1·3)

Give the absolute value of the integer. Then write its opposite.
(Lesson 1·4)

23. −8   **24.** 14   **25.** −11   **26.** 20

Add or subtract. (Lesson 1·4)

27. 9 + (−2)   **28.** 6 − 7   **29.** −6 + (−6)
30. 4 − (−4)   **31.** −7 − (−7)   **32.** −4 + 8

Compute. (Lesson 1·4)

33. −5 × (−7)   **34.** 60 ÷ (−12)
35. −48 ÷ (−8)   **36.** (−4 × 5) × (−3)
37. 2 × [−8 + (−4)]   **38.** −6 [4 − (−7)]

39. What can you say about the product of two negative integers?
(Lesson 1·4)

40. What can you say about the sum of two negative integers?
(Lesson 1·4)

HotWords

absolute value (Lesson 1·4)
Associative Property
(Lesson 1·1)
common factor (Lesson 1·3)
Commutative Property
(Lesson 1·1)
composite number (Lesson 1·3)
Distributive Property
(Lesson 1·1)
divisible (Lesson 1·3)
exponent (Lesson 1·3)
factor (Lesson 1·3)

greatest common factor
(Lesson 1·3)
least common multiple
(Lesson 1·3)
multiple (Lesson 1·3)
negative integer (Lesson 1·4)
negative number (Lesson 1·4)
operation (Lesson 1·2)
PEMDAS (Lesson 1·2)
positive integer (Lesson 1·4)
prime factorization (Lesson 1·3)
prime number (Lesson 1·3)
Venn diagram (Lesson 1·3)

1·1 Properties

Commutative and Associative Properties

The operations of addition and multiplication share special properties because multiplication is a shortcut notation for repeated addition.

Both addition and multiplication are **commutative**. This means that changing the order of the numbers does not change the sum or the product.

$$6 + 3 = 3 + 6 \text{ and } 6 \times 3 = 3 \times 6$$

If a and b are any whole numbers, then

$$a + b = b + a \text{ and } a \times b = b \times a.$$

Addition and multiplication are also **associative**. This means that changing the grouping of the addends or factors does not change the sum or the product.

$$(5 + 7) + 9 = 5 + (7 + 9) \text{ and } (3 \times 2) \times 4 = 3 \times (2 \times 4)$$
$$(a + b) + c = a + (b + c) \text{ and } (a \times b) \times c = a \times (b \times c)$$

Subtraction and division are not commutative. For example:

$6 - 4 = 2$, but $4 - 6 = -2$; therefore, $6 - 4 \neq 4 - 6$.
$6 \div 4 = 1.5$, but $4 \div 6$ is about 0.67;
therefore, $6 \div 4 \neq 4 \div 6$.

Similarly, the following examples show that subtraction and division are not associative.

$(6 - 8) - 5 = -7$, but $6 - (8 - 5) = 3$;
therefore, $(6 - 8) - 5 \neq 6 - (8 - 5)$.
$(6 \div 3) \div 4 = 0.5$, but $6 \div (3 \div 4) = 8$;
therefore, $(6 \div 3) \div 4 \neq 6 \div (3 \div 4)$.

Check It Out

Write *yes* or *no*.

1 $3 \times 4 = 4 \times 3$

2 $12 - 5 = 5 - 12$

3 $(12 \div 6) \div 3 = 12 \div (6 \div 3)$

4 $5 + (6 + 7) = (5 + 6) + 7$

Properties of One and Zero

When you add 0 to any number, the sum is always equal to the number to which zero is being added. This is called the *Identity Property of Addition*. For example:

$3 + 0 = 3$

$54 + 0 = 54$

$769 + 0 = 769$

The sum is always equal to the number that is being added to zero.

When you multiply any number by 1, the product is that number. This is called the *Identity Property of Multiplication*. For example:

$3 \times 1 = 3$

$1{,}987 \times 1 = 1{,}987$

$200{,}453 \times 1 = 200{,}453$

The product of a number and 1 is always that number.

The product of any number and 0 is 0. This is called the *Zero Property of Multiplication*. For example:

$13 \times 0 = 0$

$49 \times 0 = 0$

$158{,}975 \times 0 = 0$

The product of a number and 0 is always 0.

Check It Out

Solve.

5 $26{,}307 \times 1$

6 $199 + 0$

7 $7 \times (8 \times 0)$

8 $(4 \times 0.6) \times 1$

Distributive Property

The **Distributive Property** is important because it combines both addition and multiplication. This property states that multiplying a sum by a number is the same as multiplying each addend by that number and then adding the two products.

$$4(1 + 5) = (4 \times 1) + (4 \times 5)$$

If a, b, and c are any whole numbers, then
$$a(b + c) = ab + ac.$$

 Check It Out

Rewrite each expression, using the Distributive Property.

9 $3 \times (2 + 6)$

10 $(6 \times 7) + (6 \times 8)$

Shortcuts for Adding and Multiplying

You can use the properties to help you perform some computations mentally.

$$41 + 56 + 23 = (41 + 23) + 56 = 64 + 56 = 120$$

Use the Commutative
and Associative Properties.

$$8 \times 9 \times 30 = (8 \times 30) \times 9 = 240 \times 9 = 2,160$$

$$3 \times 220 = (3 \times 200) + (3 \times 20) = 600 + 60 = 660$$

Use the Distributive Property.

1·1 Exercises

Write *yes* or *no*.

1. $7 \times 31 = 31 \times 7$
2. $3 \times 5 \times 6 = 3 \times 6 \times 5$
3. $4 \times 120 = (3 \times 100) \times (4 \times 20)$
4. $b \times (w + p) = bw + bp$
5. $(4 \times 6 \times 5) = (4 \times 6) + (4 \times 5)$
6. $b \times (c + d + e) = bc + bd + be$
7. $13 - 8 = 8 - 13$
8. $12 \div 4 = 4 \div 12$

Solve.

9. $42{,}750 \times 1$
10. $588 + 0$
11. $6 \times (0 \times 5)$
12. $0 \times 4 \times 16$
13. 1×0
14. $3.8 + 0$
15. 5.24×1
16. $(4 + 6 + 3) \times 1$

Rewrite each expression with the Distributive Property.

17. $5 \times (7 + 4)$
18. $(8 \times 15) + (8 \times 6)$
19. 4×550

Solve using mental math.

20. $5 \times (24 + 6)$
21. $8 \times (22 + 78)$
22. 9×320
23. 15×8
24. $12 + 83 + 88$
25. $250 + 150 + 450$
26. 130×7
27. $11 \times 50 \times 2$

28. Give an example to show that subtraction is not associative.
29. Give an example to show that division is not commutative.
30. How would you describe the Identity Property of Addition?

1·2 Order of Operations

Understanding the Order of Operations

Solving a problem may involve using more than one **operation**. Your answer will depend on the order in which you complete those operations.

For example, consider the expression $2^2 + 5 \times 6$.

$$2^2 + 5 \times 6 \qquad \text{or} \qquad 2^2 + 5 \times 6$$

$$4 + 5 \times 6 \qquad\qquad\quad 4 + 5 \times 6$$

$$9 \times 6 = \boxed{54} \qquad\qquad 4 + 30 = \boxed{34}$$

The order in which you perform operations makes a difference.

To make sure that there is just one answer to a series of computations, mathematicians have agreed upon an order in which to do the operations.

EXAMPLE   **Using the Order of Operations**

Simplify $4^2 - 8 \times (6 - 6)$.

$= 4^2 - 8 \times (0)$ • Simplify within parentheses.

$= (16) - 8 \times 0$ • Evaluate the power.

$= 16 - 0$ • Multiply or divide from left to right.

$= 16$ • Add or subtract from left to right.

So, $4^2 - 8 \times (6 - 6) = 16$.

 Check It Out

Simplify.

1 $20 - 2 \times 3$ **2** $3 \times (2 + 5^2)$

3 $5^2 + 3 \times (1 + 4)$ **4** $8 \times (2 - 1) + 6 \times 3$

1·2 Exercises

Is each expression true? Write *yes* or *no*.

1. $7 \times 3 + 5 = 26$

2. $3 + 5 \times 7 = 56$

3. $6 \times (8 + 4 \div 2) = 36$

4. $6^2 - 1 = 25$

5. $(1 + 7)^2 = 64$

6. $(2^3 + 5 \times 2) + 6 = 32$

7. $45 - 5^2 = 20$

8. $(4^2 \div 4)^3 = 64$

Simplify.

9. $24 - (3 \times 5)$

10. $3 \times (4 + 5^2)$

11. $2^4 \times (10 - 7)$

12. $4^2 + (4 - 3)^2$

13. $(14 - 10)^2 \times 6$

14. $12 + 9 \times 3^2$

15. $(3^2 + 3)^2$

16. $48 \div (12 + 4)$

17. $30 - (10 - 6)^2$

18. $34 + 6 \times (4^2 \div 2)$

Use parentheses to make the expression true.

19. $5 + 5 \times 6 = 60$

20. $4 \times 25 + 75 = 400$

21. $36 \div 6 + 3 = 4$

22. $20 + 20 \div 4 - 4 = 21$

23. $10 \times 3^2 + 5 = 140$

24. $5^2 - 12 \div 3 \times 2^2 = 84$

25. Use five 2s, a set of parentheses (as needed), and any of the operations to make the numbers 1 through 3.

(**P** arentheses)

E xponents ^{2}

✗ **M** ultiplication &

D ivision ÷

✚ **A** ddition &

S ubtraction —

1·3 Factors and Multiples

Factors

Suppose that you want to arrange 18 small squares in a rectangular pattern. All possible patterns are shown below.

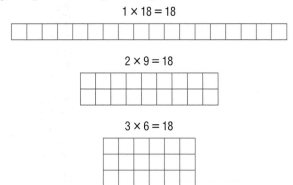

1 × 18 = 18

2 × 9 = 18

3 × 6 = 18

Two numbers multiplied together to produce 18 are considered **factors** of 18. So, the factors of 18 are 1, 2, 3, 6, 9, and 18.

To decide whether one number is a factor of another, divide. If there is a remainder of 0, the number is a factor.

EXAMPLE **Finding the Factors of a Number**

What are the factors of 28?

• Find all pairs of numbers that multiply to the product.

   1 × 28 = 28   2 × 14 = 28   4 × 7 = 28

• List the factors in order, starting with 1.

The factors of 28 are 1, 2, 4, 7, 14, and 28.

 Check It Out

Find the factors of each number.

 15

 16

Common Factors

Factors that are the same for two or more numbers are called **common factors**.

EXAMPLE Finding Common Factors

What numbers are factors of both 12 and 30?

1, 2, 3, 4, 6, 12	• List the factors of the first number, 12.
1, 2, 3, 5, 6, 10, 15, 30	• List the factors of the second number, 30.
1, 2, 3, 6	• The common factors are the numbers that are in both lists.

The common factors of 12 and 30 are 1, 2, 3, and 6.

 Check It Out

List the common factors of each set of numbers.

3 8 and 20   **4** 14, 21, 35

Greatest Common Factor

The **greatest common factor** (GCF) of two whole numbers is the greatest number that is a factor of both numbers.

One way to find the GCF is to list the common factors and choose the greatest common factor.

What is the GCF of 16 and 40?
• The factors of 16 are 1, ②, ④, ⑧, 16.
• The factors of 40 are 1, ②, ④, 5, ⑧, 10, 20, 40.
• The common factors that are in both lists are 1, 2, 4, 8.
The GCF of 16 and 40 is 8.

 Check It Out

Find the GCF for each pair of numbers.

5 8 and 20   **6** 10 and 60

Venn Diagram

A **Venn diagram** uses circles to show how the elements of two or more sets are related. When the circles in a Venn diagram overlap, the overlapping part contains elements that are common to both sets.

A Venn diagram can be used to find the common factors of 10 and 36. The elements of set A are all factors of 10. The elements of set B are all factors of 36. The overlapping part of A and B contains the elements that are factors of both 10 and 36.

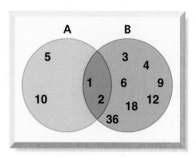

The common factors of 10 and 36 are 1 and 2.

A Venn diagram with three circles can be used to find the common factors of three numbers. When there are more complex Venn diagrams, you have to look carefully at the overlapping parts to identify which elements of the sets are in those parts.

The common factors of 5, 10, and 25 are 1 and 5.

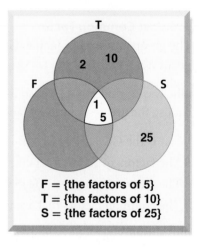

F = {the factors of 5}
T = {the factors of 10}
S = {the factors of 25}

 Check It Out

Use a Venn diagram to find the common factors for each set of numbers.

7 3 and 15

8 2, 4, and 32

Divisibility Rules

There are times when you will want to know whether a number is a factor of a much larger number. For instance, if you want to share 231 CDs among 3 friends, you will need to know whether 231 is divisible by 3. A number is **divisible** by another number if the remainder of their quotient is zero.

> You can quickly figure out whether 231 is divisible by 3 if you know the divisibility rule for 3. A number is divisible by 3 if the sum of the digits is divisible by 3. For example, 231 is divisible by 3 because $2 + 3 + 1 = 6$, and 6 is divisible by 3.

It can be helpful to know other divisibility rules. A number is divisible by

2 if the ones digit is 0 or an even number.

3 if the sum of the digits is divisible by 3.

4 if the number formed by the last two digits is divisible by 4.

5 if the ones digit is 0 or 5.

6 if the number is divisible by 2 and 3.

8 if the number formed by the last three digits is divisible by 8.

9 if the sum of the digits is divisible by 9.

And . . .

Any number is divisible by **10** if the ones digit is 0.

 Check It Out

Use the divisibility rules to solve.

9 Is 536 divisible by 4?

10 Is 929 divisible by 9?

11 Is 626 divisible by 6?

12 Is 550 divisible by 5?

Prime and Composite Numbers

A **prime number** is a whole number greater than 1 with exactly two factors, itself and 1. Here are the first 10 prime numbers:

2, 3, 5, 7, 11, 13, 17, 19, 23, 29

Twin primes are pairs of primes that have a difference of 2. (3, 5), (5, 7), and (11, 13) are examples of twin primes.

A number with more than two factors is called a **composite number**. When two composite numbers have no common factors (other than 1), they are said to be *relatively prime.* The numbers 8 and 15 are relatively prime.

Identifying Prime and Composite Numbers		
Example	**Factors**	**Classification**
$16 = 1 \times 16$ $16 = 2 \times 8$ $16 = 4 \times 4$	5 factors	composite
$13 = 1 \times 13$	exactly 2 factors	prime

Identifying Relative Prime Numbers		
Example	**Factors**	**Classification**
$9 = 1 \times 9$ $9 = 3 \times 3$	①, 3, 9	9 and 14 are composite numbers that are relatively prime because the only common factor is 1.
$14 = 1 \times 14$ $14 = 2 \times 7$	①, 2, 7, 14	

 Check It Out

Write whether the following numbers are prime.

13 71

14 87

15 97

16 106

Prime Factorization

Every composite number can be expressed as a product of prime factors. Use a factor tree to find the prime factors.

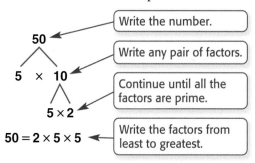

Write the number.

Write any pair of factors.

Continue until all the factors are prime.

$50 = 2 \times 5 \times 5$

Write the factors from least to greatest.

Although the order of the factors may be different because you can start with different pairs of factors, every factor tree for 50 has the same **prime factorization**. You can also use **exponents** to write the prime factorization: $50 = 2 \times 5^2$.

 Check It Out

What is the prime factorization of each number?

 17 40

18 100

Shortcut to Finding GCF

Use prime factorization to find the greatest common factor.

EXAMPLE Using Prime Factorization to Find the GCF

Find the greatest common factor of 12 and 18.

$12 = 2 \times 2 \times 3$	• Find the prime factorization of each
$18 = 2 \times 3 \times 3$	number. Use a factor tree if necessary.
2 and 3	• Find the prime factors common to both numbers.
$2 \times 3 = 6$	• Find their product.

The greatest common factor of 12 and 18 is 2×3, or 6.

Check It Out

Use prime factorization to find the GCF of each pair of numbers.

19 6 and 20   **20** 10 and 35   **21** 105 and 42

Multiples and Least Common Multiples

The **multiples** of a number are the whole-number products resulting when that number is a factor. In other words, you can find a multiple of a number by multiplying it by 1, 2, 3, and so on.

The **least common multiple** (LCM) of two numbers is the smallest positive number that is a multiple of both. One way to find the LCM of a pair of numbers is first to list multiples of each and then to identify the smallest one common to both. For instance, to find the LCM of 6 and 9:

• List multiples of 6: 6, 12, (18,) 24, 30, . . .
• List multiples of 9: 9, (18,) 27, 36, . . .
• LCM = 18

Another way to find the LCM is to use prime factorization.

EXAMPLE   Using Prime Factorization to Find the LCM

Find the least common multiple of 6 and 9.

• Find the prime factors of each number.

    $6 = 2 \times 3$ $9 = 3 \times 3$

• Multiply the prime factors of the least number by the prime factors of the greatest number that are not factors of the least number.

    $2 \times 3 \times 3 = 18$

The least common multiple of 6 and 9 is 18.

Check It Out

Use either method to find the LCM of each pair of numbers.

22 6 and 8   **23** 12 and 40   **24** 8 and 18

In the game of darts, players alternate turns, each tossing three darts at a round board. The board is divided into 20 wedges, numbered randomly from 1 to 20. You score the value of the wedge where your dart lands. The board has three rings. The bull's-eye is worth 50 points; the ring around the bull's-eye is worth 25 points; the second ring triples the value of the wedge; and the third ring doubles the value of the wedge.

Say in the first round that you throw three darts. One lands in the regular 4 space; the second lands in the second ring in the 15 wedge; and the third lands in the ring next to the bull's-eye. You earn 74 points.

Scoring works backward, by subtracting the number of points from a target number, such as 1,001 or 2,001. The record for the fewest number of darts thrown for a score of 1,001 is 19, held by Cliff Inglis (1975) and Jocky Wilson (1989). Inglis threw scores of 160, 180, 140, 180, 121, 180, and 40. Remember that Inglis's first six scores were made by tossing three darts and the last score was made by tossing one dart. What is one possible way that Inglis could have thrown his record game? See **HotSolutions** for answer.

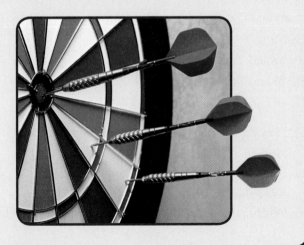

1·3 Exercises

Find the factors of each number.

1. 15

2. 24

3. 32

4. 56

Write whether the following numbers are prime.

5. 81

6. 97

7. 107

8. 207

Write the prime factorization for each number.

9. 60

10. 120

11. 160

12. 300

Find the GCF for each pair of numbers.

13. 12 and 24

14. 8 and 30

15. 18 and 45

16. 20 and 35

17. 16 and 40

18. 15 and 42

Find the LCM for each pair of numbers.

19. 6 and 7

20. 12 and 24

21. 16 and 24

22. 10 and 35

23. How do you use prime factorization to find the GCF of two numbers?

24. A mystery number is a common multiple of 2, 4, 5, and 15. It is a factor of 120. What is the number?

25. What is the divisibility rule for 8? Is 4,128 divisible by 8?

1·4 Integer Operations

Positive and Negative Integers

A glance through any newspaper shows that many quantities are expressed with **negative numbers**. For example, negative numbers show below-zero temperatures, drops in the value of stocks, or business losses.

Whole numbers greater than zero are called **positive integers**. Whole numbers less than zero are called **negative integers**.

Here is the set of all integers:
$$\{\ldots, -5, -4, -3, -2, -1, 0, 1, 2, 3, 4, 5, \ldots\}$$

The integer 0 is neither positive nor negative. A number without a sign is assumed to be a positive number.

 Check It Out
Write an integer to describe the situation.
1 4 below zero   **2** a gain of $300

Opposites of Integers and Absolute Value

Integers can describe opposite ideas. Each integer has an opposite.
　　The opposite of +4 is −4.
　　The opposite of spending $5 is earning $5.
　　The opposite of −5 is +5.
The **absolute value** of an integer is its distance from 0 on the number line. You write the absolute value of −3 as |−3|.

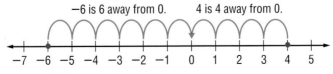

−6 is 6 away from 0.　　4 is 4 away from 0.

The absolute value of 4 is 4. You write |4| = 4.
The absolute value of −6 is 6. You write |−6| = 6.

Check It Out

Give the absolute value of the integer. Then write the opposite of the original integer.

3 −15 **4** +3 **5** −12

Comparing and Ordering Integers

When two numbers are graphed on a number line, the number to the left is always less than the number to the right. Similarly, the number to the right is always greater than the number to the left.

A negative number is always less than a positive number.

−2 is less than 2, so −2 < 2.

 Check It Out

Replace □ with < or > to make a true sentence.

6 −5 □ −6 **7** −2 □ −4 **8** −9 □ 0

If you can compare integers, then you can order a group of integers.

EXAMPLE Ordering Integers

A science class collected the temperature data shown below. Order the daily temperatures from coldest to warmest.

$$-5°F, -2°F, 8°F, 5°F, -3°F$$

• To order the integers, graph them on a number line.

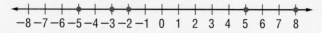

• Order the integers by reading from left to right.

−5, −3, −2, 5, and 8

The daily temperatures from coldest to warmest are −5°F, −3°F, −2°F, 5°F, and 8°F.

Order the integers in each set from least to greatest.

9 {−13, 8, −2, 0, 6}   **10** {6, −16, −10, 19, 18}

Adding and Subtracting Integers

You can use a number line to model adding and subtracting integers.

EXAMPLE Adding and Subtracting Integers

Solve 2 + (−3).

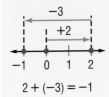

$$2 + (−3) = −1$$

Solve 6 − 4.

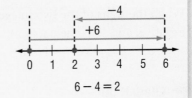

$$6 − 4 = 2$$

Solve −2 + (−4).

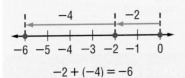

$$−2 + (−4) = −6$$

Solve 3 − 5.

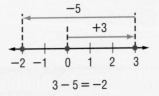

$$3 − 5 = −2$$

Rules for Adding or Subtracting Integers		
To	**Solve**	**Example**
Add integers with the same sign	Add the absolute values. Use the original sign in answer.	−3 + (−3): $\lvert −3 \rvert + \lvert −3 \rvert = 3 + 3 = 6$ So, −3 + (−3) = −6.
Add integers with different signs	Subtract the absolute values. Use the sign of addend with greater absolute value in answer.	−6 + 4: $\lvert −6 \rvert − \lvert 4 \rvert = 6 − 4 = 2$ $\lvert −6 \rvert > \lvert 4 \rvert$ So, −6 + 4 = −2.
Subtract integers	Add the opposite.	−4 − 2 = −4 + (−2) = −6

Check It Out

Solve.

⑪ $5 - 8$ ⑫ $7 + (-7)$ ⑬ $-8 - (-6)$ ⑭ $0 + (-4)$

Multiplying and Dividing Integers

Multiply and divide integers as you would whole numbers. Then use these rules for writing the sign of the answer.

The product of two integers with like signs is positive, and so is the quotient.

$$-5 \times (-3) = 15 \qquad\qquad -18 \div (-6) = 3$$

When the signs of the two integers are different, the product is negative, and so is the quotient.

$$-6 \div 2 = -3 \qquad -4 \times 6 = -24 \qquad -5 \times 10 = -50$$

 Check It Out

Find the product or quotient.

⑮ $-2 \times (-4)$ ⑯ $12 \div (-3)$

⑰ $-16 \div (-4)$ ⑱ -7×8

APPLICATION **Double Your Fun or Not**

Did you know you can use integer math to help you with English grammar? What is the opposite of *not inside*?

When you use two negatives in math, such as $-(-3)$, you are really asking, "What is the opposite of the number inside the parentheses?" Since the opposite of -3 is $+3$, $-(-3) = 3$. Now try the original question again—the opposite of *not inside* is *not (not inside)*, or *inside*! So, the mathematical idea that two negatives make a positive applies to English grammar as well.

1·4 Exercises

Give the absolute value of the integer. Then write its opposite.

1. -13
2. 7
3. -5
4. 1

Add or subtract.

5. $5 - 4$
6. $4 + (-7)$
7. $-7 - (-6)$
8. $0 + (-6)$
9. $-2 + 8$
10. $0 - 9$
11. $0 - (-7)$
12. $-3 - 9$
13. $6 + (-6)$
14. $-8 - (-5)$
15. $-4 - (-4)$
16. $-8 + (-7)$

Find the product or quotient.

17. $-2 \times (-8)$
18. $9 \div (-3)$
19. $-25 \div 5$
20. -4×7
21. $5 \times (-9)$
22. $-32 \div 8$
23. $-15 \div (-3)$
24. $4 \times (-9)$

Compute.

25. $[-5 \times (-2)] \times 4$
26. $4 \times [3 \times (-4)]$
27. $[-3 \times (-4)] \times (-3)$
28. $-5 \times [3 + (-4)]$
29. $(-7 - 2) \times 3$
30. $-4 \times [6 - (-2)]$

31. Is the absolute value of a positive integer positive or negative?
32. If you know that the absolute value of an integer is 6, what are the possible values of that integer?
33. What can you say about the sum of two positive integers?
34. What can you say about the product of a negative integer and a positive integer?
35. The temperature at noon was 12°F. For the next 5 hours, it dropped at a rate of 2 degrees an hour. First express this change as an integer, and then give the temperature at 5 P.M.

Numbers and Computation

What have you learned?

You can use the problems and the list of words that follow to see what you learned in this chapter. You can find out more about a particular problem or word by referring to the topic number (*for example,* Lesson 1·2).

Problem Set

Solve. (Lesson 1·1)

1. 836×0 **2.** $(5 \times 3) \times 1$ **3.** $6{,}943 + 0$ **4.** 0×0

Solve using mental math. (Lesson 1·1)

5. $3 \times (34 + 66)$ **6.** $(50 \times 12) \times 2$

Use parentheses to make each expression true. (Lesson 1·2)

7. $4 + 8 \times 2 = 24$ **8.** $35 + 12 \div 2 + 5 = 46$

Is each expression true? Write *yes* or *no*. (Lesson 1·2)

9. $4 \times 4^2 = 64$ **10.** $2^3 \times (7 + 5 \div 3) = 36$

Simplify. (Lesson 1·2)

11. $6^2 - (3 \times 6)$ **12.** $5 \times (8 + 5^2)$ **13.** $2^4 \times (10 - 3)$

Is it a prime number? Write *yes* or *no*. (Lesson 1·3)

14. 49 **15.** 105 **16.** 163 **17.** 203

Write the prime factorization for each number. (Lesson 1·3)

18. 25 **19.** 170 **20.** 300

Find the GCF for each pair of numbers. (Lesson 1·3)

21. 16 and 30   **22.** 12 and 50   **23.** 10 and 160

Find the LCM for each pair of numbers. (Lesson 1·3)

24. 5 and 12   **25.** 15 and 8   **26.** 18 and 30

27. What is the divisibilty rule for 5? Is 255 a multiple of 5?
(Lesson 1·3)

Give the absolute value of the integer. Then write its opposite. (Lesson 1·4)

28. -3 **29.** 16   **30.** -12 **31.** 25

Add or subract. (Lesson 1·4)

32. $10 + (-8)$ **33.** $7 - 8$ **34.** $-4 + (-5)$
35. $6 - (-6)$ **36.** $-9 - (-9)$ **37.** $-6 + 14$

Compute. (Lesson 1·5)

38. $-9 \times (-9)$ **39.** $48 \div (-12)$ **40.** $-27 \div (-9)$
41. $(-4 \times 3) \times (-4)$ **42.** $4 \times [-5 + (-7)]$ **43.** $-4[5 - (-9)]$

44. What can you say about the quotient of two negative integers? (Lesson 1·4)

45. What can you say about the difference of two positive integers? (Lesson 1·4)

HotWords

Write definitions for the following words.

absolute value (Lesson 1·4)
Associative Property
    (Lesson 1·1)
common factor (Lesson 1·3)
Commutative Property
    (Lesson 1·1)
composite number (Lesson 1·3)
Distributive Property
    (Lesson 1·1)
divisible (Lesson 1·3)
exponent (Lesson 1·3)
factor (Lesson 1·3)

greatest common factor
    (Lesson 1·3)
least common multiple
    (Lesson 1·3)
multiple (Lesson 1·3)
negative integer (Lesson 1·4)
negative number (Lesson 1·4)
operation (Lesson 1·2)
PEMDAS (Lesson 1·2)
positive integer (Lesson 1·4)
prime factorization (Lesson 1·3)
prime number (Lesson 1·3)
Venn diagram (Lesson 1·3)

HotTopic 2

Fractions, Decimals, and Percents

What do you know?

You can use the problems and the list of words that follow to see what you already know about this chapter. The answers to the problems are in **HotSolutions** at the back of the book, and the definitions of the words are in **HotWords** at the front of the book. You can find out more about a particular problem or word by referring to the topic number (*for example*, Lesson 2·2).

Problem Set

1. Mr. Sebo is on a special diet. He must carefully weigh and measure his food. He can eat 12.25 oz at breakfast, 14.621 oz at lunch, and 20.03 oz at dinner. How many ounces of food is he allowed to eat each day? (Lesson 2·6)

2. Ying got 2 out of 25 problems wrong on her math quiz. What percent did she get correct? (Lesson 2·8)

3. Which fraction is not equivalent to $\frac{3}{12}$? (Lesson 2·1)

   A. $\frac{12}{48}$ B. $\frac{1}{4}$ C. $\frac{18}{60}$ D. $\frac{24}{96}$

Add or subtract. (Lesson 2·3)

4. $\frac{4}{5} + \frac{1}{6}$

5. $3\frac{2}{7} - 2\frac{1}{7}$

6. Find the improper fraction, and write it as a mixed number. (Lesson 2·1)

   A. $\frac{5}{12}$ B. $\frac{3}{2}$ C. $2\frac{1}{2}$ D. $\frac{18}{36}$

Multiply or divide. (Lesson 2·4)

7. $\frac{3}{4} \times \frac{2}{5}$

8. $6\frac{1}{2} \div 3\frac{1}{2}$

9. Write 2.002 in expanded form. (Lesson 2·5)

10. Write as a decimal: three hundred and three hundred three thousandths. (Lesson 2·5)

Solve. (Lesson 2·6)

11. $2{,}504 + 11.66$

12. $10.5 - 9.06$

13. 3.25×4.1

14. $41.76 \div 1.2$

Use a calculator. Round to the nearest tenth. (Lesson 2·8)

15. What percent of 48 is 12?

16. Find 4% of 50.

Write each decimal or fraction as a percent. (Lesson 2·9)

17. 0.99   **18.** 0.4   **19.** $\dfrac{7}{100}$ **20.** $\dfrac{83}{100}$

Write each percent as a fraction in simplest form. (Lesson 2·9)

21. 27%   **22.** 120%

Order these rational numbers from least to greatest. (Lesson 2·9)

23. $0.7, 7\%, \dfrac{3}{4}$

benchmark (Lesson 2·7)

common denominator
(Lesson 2·1)

compatible numbers
(Lesson 2·6)

cross product (Lesson 2·1)

denominator (Lesson 2·1)

discount (Lesson 2·8)

equivalent (Lesson 2·1)

equivalent fractions (Lesson 2·1)

estimate (Lesson 2·6)

factor (Lesson 2·4)

fraction (Lesson 2·1)

greatest common factor
(Lesson 2·1)

improper fraction (Lesson 2·1)

least common multiple
(Lesson 2·1)

mixed number (Lesson 2·1)

numerator (Lesson 2·1)

percent (Lesson 2·7)

product (Lesson 2·4)

ratio (Lesson 2·7)

rational number (Lesson 2·9)

reciprocal (Lesson 2·4)

repeating decimal (Lesson 2·9)

terminating decimal
(Lesson 2·9)

2·1 Fractions and Equivalent Fractions

Naming Fractions

A **fraction** can be used to name a part of a whole. For example, the flag of Bolivia is divided into three equal parts: red, yellow, and green. Each part, or color, of the Bolivian flag represents $\frac{1}{3}$ of the whole flag. $\frac{3}{3}$, or 1, represents the whole flag.

A fraction can also name part of a set.

There are six balls in the set of balls. Each ball is $\frac{1}{6}$ of the set. $\frac{6}{6}$, or 1, equals the whole set. Five of the balls are soccer balls. The soccer balls represent $\frac{5}{6}$ of the set. One of the six balls is a basketball. The basketball represents $\frac{1}{6}$ of the set.

You use **numerators** and **denominators** to name fractions.

Write a fraction for the number of shaded squares.

There are 8 squares in all.

5 squares are shaded.

- The denominator of the fraction tells the total number of parts.

- The numerator of the fraction tells the number of parts being considered.

$$\frac{\text{parts being considered}}{\text{parts that make a whole}} = \frac{\text{numerator}}{\text{denominator}}$$

- Write the fraction.

$\frac{5}{8}$ is the fraction for the number of shaded squares.

Check It Out

Write the fraction for each picture.

1 ____ of the circle is shaded.

2 ____ of the triangles are shaded.

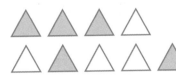

3 Draw two pictures to represent the fraction $\frac{2}{5}$. Use regions and sets.

Methods for Finding Equivalent Fractions

Equivalent fractions are fractions that describe the same amount of a region. You can use fraction pieces to show equivalent fractions.

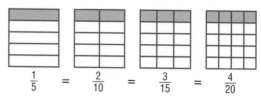

$$\frac{1}{5} = \frac{2}{10} = \frac{3}{15} = \frac{4}{20}$$

Each of the fraction pieces represents a fraction equal to $\frac{1}{5}$.

Fraction Names for One

There is an infinite number of fractions that are equal to one.

Names for one

$$\frac{4}{4} \qquad \frac{13}{13} \qquad \frac{1}{1} \qquad \frac{3{,}523}{3{,}523}$$

Any number multiplied by 1 is still equal to the original number. So, knowing different fraction names for 1 can help you find equivalent fractions.

You can find a fraction that is equivalent to another fraction by multiplying the fraction by a form of 1 or by dividing the numerator and the denominator by the same number.

EXAMPLE Finding Equivalent Fractions

Find a fraction equal to $\frac{4}{8}$.

Multiply by a form of one.

$$\frac{4}{8} \times \frac{3}{3} = \frac{12}{24} \qquad \frac{4}{8} = \frac{12}{24}$$

$$\frac{4 \div 2}{8 \div 2} = \frac{2}{4} \qquad \frac{4}{8} = \frac{2}{4}$$

Divide the numerator and the denominator by the same number.

Write two fractions equivalent to each fraction.

4 $\frac{1}{2}$

5 $\frac{12}{24}$

6 Write three fraction names for one.

Deciding Whether Two Fractions Are Equivalent

Two fractions are **equivalent** if you can show that each fraction is just a different name for the same amount.

The fraction piece shows $\frac{1}{3}$ of the whole circle.

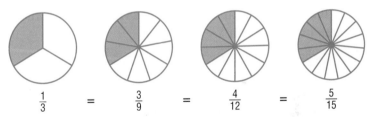

$$\frac{1}{3} \quad = \quad \frac{3}{9} \quad = \quad \frac{4}{12} \quad = \quad \frac{5}{15}$$

You can use fraction pieces to show names for the same amount.

You can identify equivalent fractions by comparing their **cross products**.

> **EXAMPLE**   **Deciding Whether Two Fractions Are Equivalent**
>
> Determine if $\frac{1}{2}$ and $\frac{4}{8}$ are equivalent fractions.
>
>   • Find the cross products of the fractions.
>
> $1 \times 8 \overset{?}{=} 4 \times 2$
>
> $8 \overset{?}{=} 8$
>
> $8 = 8$ • Compare the cross products.
>
>   • If the cross products are equal, then the fractions are equivalent.
>
> So, $\frac{1}{2} = \frac{4}{8}$.

 Check It Out

Use the cross products method to determine whether the fractions in each pair are equivalent.

7 $\frac{3}{8}, \frac{9}{24}$ **8** $\frac{8}{12}, \frac{4}{6}$ **9** $\frac{20}{36}, \frac{5}{8}$

Least Common Denominator

Fractions that have the same denominators are called *like fractions*. Fractions that have different denominators are called *unlike fractions*. $\frac{1}{2}$ and $\frac{1}{4}$ are unlike fractions.

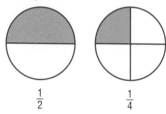

$\frac{1}{2}$ $\frac{1}{4}$

FRACTIONS AND EQUIVALENT FRACTIONS   **2·1**

Fractions can be renamed as like fractions. A common denominator is a common multiple of the denominators of fractions. The *least common denominator* (LCD) is the *least common multiple* (LCM) of the denominator.

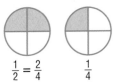

$$\frac{1}{2} = \frac{2}{4} \qquad \frac{1}{4}$$

EXAMPLE **Finding the Least Common Denominator**

Find the least common denominator (LCD) for the fractions $\frac{2}{3}$ and $\frac{3}{5}$.

- Find the least common multiple (LCM) (p. 82) of the denominators.

  3: 3, 6, 9, 12, (15,) 18, 21, 24, 27

  5: 5, 10, (15,) 20, 25, 30, 35, 40, 45

  15 is the LCD for the fractions $\frac{2}{3}$ and $\frac{3}{5}$. It is the least common multiple of the denominators.

- To find equivalent fractions with the LCD, find the right form of 1 to multiply with each fraction.

$$\frac{2}{3} \times \frac{?}{?} = \frac{}{15} \qquad\qquad \frac{3}{5} \times \frac{?}{?} = \frac{}{15}$$

$3 \times 5 = 15$, so use $\frac{5}{5}$. $\qquad 5 \times 3 = 15$, so use $\frac{3}{3}$.

$$\frac{2}{3} \times \frac{5}{5} = \frac{10}{15} \qquad\qquad \frac{3}{5} \times \frac{3}{3} = \frac{9}{15}$$

Fractions $\frac{2}{3}$ and $\frac{3}{5}$ become $\frac{10}{15}$ and $\frac{9}{15}$ when renamed with their least common denominator.

 Check It Out

Find the LCD. Write like fractions.

⑩ $\frac{3}{8}, \frac{1}{4}$ ⑪ $\frac{7}{10}, \frac{21}{50}$

⑫ $\frac{3}{10}, \frac{1}{2}$ ⑬ $\frac{7}{12}, \frac{8}{18}$

Writing Fractions in Simplest Form

A fraction is in simplest form if the greatest common factor (GCF) of the numerator and the denominator is 1.

These fractions are all equivalent to $\frac{3}{12}$:

$$\frac{18}{72} \qquad \frac{12}{48} \qquad \frac{6}{24} \qquad \frac{3}{12} \qquad \frac{2}{8} \qquad \frac{1}{4}$$

$$\frac{3}{12} = \frac{1}{4}$$

The smallest number of fraction pieces that show a fraction equivalent to $\frac{3}{12}$ is $\frac{1}{4}$. The fraction $\frac{3}{12}$ expressed in simplest form is $\frac{1}{4}$. Another way to find the simplest form is to divide the numerator and denominator by the **greatest common factor**.

EXAMPLE Finding Simplest Form of Fractions

Express $\frac{12}{30}$ in simplest form.

• Find the greatest common factor (p. 77) of the numerator and denominator.

    12: 1, 2, 3, 4, 6, 12

    30: 1, 2, 3, 5, 6, 10, 15, 30

    The GCF is 6.

• Divide the numerator and the denominator of the fraction by the GCF.

$$\frac{12 \div 6}{30 \div 6} = \frac{2}{5}$$

$\frac{2}{5}$ is $\frac{12}{30}$ in simplest form.

Check It Out

Express each fraction in simplest form.

14 $\frac{2}{14}$ **15** $\frac{7}{28}$ **16** $\frac{25}{30}$

Writing Improper Fractions and Mixed Numbers

You can express amounts greater than 1 as *improper fractions*. A fraction with a numerator greater than or equal to the denominator is called an **improper fraction**. $\frac{9}{4}$ is an example of an improper fraction. Another name for $\frac{9}{4}$ is $2\frac{1}{4}$. A whole number and a fraction make up a **mixed number**, so $2\frac{1}{4}$ is a mixed number.

You can write any mixed number as an improper fraction and any improper fraction as a mixed number.

EXAMPLE Changing an Improper Fraction to a Mixed Number

Change $\frac{15}{6}$ to a mixed number.

- Divide the numerator by the denominator.

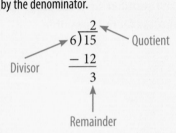

- Write the mixed number.

- Express in simplest form.
$$\frac{3}{6} = \frac{3 \div 3}{6 \div 3} = \frac{1}{2}$$
So, $\frac{15}{6} = 2\frac{1}{2}$.

You can use multiplication to write a mixed number as an improper fraction.

EXAMPLE | **Changing a Mixed Number to an Improper Fraction**

Write $2\frac{1}{5}$ as an improper fraction.

$2 \times \frac{5}{5} = \frac{10}{5}$

- Rename the whole number part by multiplying it by a form of 1 that has the same denominator as the fraction part.

$2\frac{1}{5} = \frac{10}{5} + \frac{1}{5} = \frac{11}{5}$

- Add the two parts.

You can write the mixed number $2\frac{1}{5}$ as the improper fraction $\frac{11}{5}$.

 Check It Out

Write a mixed number for each improper fraction. Write in simplest form if possible.

17 $\frac{19}{6}$

18 $\frac{15}{9}$

19 $\frac{16}{9}$

20 $\frac{57}{8}$

Write an improper fraction for each mixed number.

21 $8\frac{3}{4}$

22 $15\frac{1}{4}$

23 $16\frac{2}{3}$

24 $5\frac{9}{10}$

2·1 Exercises

Write the fraction for each picture.

1. ____ of the pieces of fruit are oranges.

2. ____ of the circle is shaded.

3. ____ of the triangles are green.

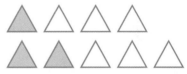

4. ____ of the balls are softballs.

Write the fraction.

5. five eighths

6. eleven fifteenths

Write one fraction equivalent to the given fraction.

7. $\frac{3}{4}$ 8. $\frac{5}{6}$ 9. $\frac{10}{50}$ 10. $\frac{8}{16}$ 11. $\frac{12}{16}$

Express each fraction in simplest form.

12. $\frac{28}{35}$ 13. $\frac{20}{36}$ 14. $\frac{66}{99}$ 15. $\frac{42}{74}$ 16. $\frac{22}{33}$

Write each improper fraction as a mixed number. Write in simplest form if possible.

17. $\frac{34}{10}$ 18. $\frac{38}{4}$ 19. $\frac{14}{3}$ 20. $\frac{44}{6}$ 21. $\frac{56}{11}$

Write each mixed number as an improper fraction.

22. $12\frac{5}{6}$ 23. $9\frac{7}{20}$ 24. $1\frac{7}{12}$ 25. $1\frac{3}{21}$

2·2 Comparing and Ordering Fractions

You can use fraction pieces to compare fractions.

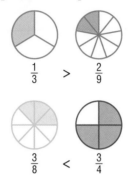

$$\frac{1}{3} \quad > \quad \frac{2}{9}$$

$$\frac{3}{8} \quad < \quad \frac{3}{4}$$

To compare fractions, you can also find *equivalent fractions* (p. 98) and compare numerators.

EXAMPLE **Comparing Fractions**

Compare the fractions $\frac{5}{6}$ and $\frac{7}{8}$.

- Look at the denominators.

$$\frac{5}{\textcircled{6}} \quad \text{and} \quad \frac{7}{\textcircled{8}}$$

Denominators
are different.

- Write equivalent fractions with a common denominator.

$$\frac{5}{6} \times \frac{4}{4} = \frac{20}{24} \qquad \frac{7}{8} \times \frac{3}{3} = \frac{21}{24}$$

- Compare the numerators.

$$20 < 21$$

- The fractions compare as the numerators compare.

$$\frac{20}{24} < \frac{21}{24}$$

So, $\frac{5}{6} < \frac{7}{8}$.

Compare the fractions. Use <, >, or =.

1 $\frac{2}{9} \square \frac{1}{4}$

2 $\frac{3}{14} \square \frac{2}{43}$

3 $\frac{3}{28} \square \frac{4}{19}$

4 $\frac{3}{13} \square \frac{1}{6}$

Comparing Mixed Numbers

To compare mixed numbers, first compare the whole numbers. Then compare the fractions, if necessary.

EXAMPLE **Comparing Mixed Numbers**

Compare $1\frac{1}{5}$ and $1\frac{2}{7}$.

- Make sure that the fraction parts are not improper.

$\frac{1}{5}$ and $\frac{2}{7}$ are not improper.

- Compare the whole-number parts. If they are different, the one that is greater is the greater mixed number. If they are equal, you must compare the fraction parts.

$$1 = 1$$

- To compare the fraction parts, rename them with a *common denominator* (p. 99).

35 is the least common multiple of 5 and 7.
Use 35 for the common denominator.

$$\frac{1}{5} \times \frac{7}{7} = \frac{7}{35} \quad \text{and} \quad \frac{2}{7} \times \frac{5}{5} = \frac{10}{35}.$$

- Compare the fractions.

$$\frac{7}{35} < \frac{10}{35}$$

So, $1\frac{1}{5} < 1\frac{2}{7}$.

 Check It Out

Compare the mixed numbers. Use <, >, or =.

5 $3\frac{1}{2} \square 3\frac{5}{8}$

6 $2\frac{3}{7} \square 1\frac{5}{12}$

7 $2\frac{4}{9} \square 3\frac{5}{11}$

Ordering Fractions

To compare and order fractions, you can find equivalent fractions and then compare the numerators of the fractions.

EXAMPLE Ordering Fractions with Unlike Denominators

Order the fractions $\frac{3}{5}$, $\frac{3}{4}$, and $\frac{7}{10}$ from least to greatest.

• Find the *least common denominator* (LCD) (p. 82) for the fractions.

Multiples of

4: 4, 8, 12, 16, (20) 24, . . .

5: 5, 10, 15, (20) 25, . . .

10: 10, (20) 30, 40, . . .

20 is the LCD of 4, 5, and 10.

• Write equivalent fractions with the least common denominator.

$$\frac{3}{5} = \frac{3}{5} \times \frac{4}{4} = \frac{12}{20}$$

$$\frac{3}{4} = \frac{3}{4} \times \frac{5}{5} = \frac{15}{20}$$

$$\frac{7}{10} = \frac{7}{10} \times \frac{2}{2} = \frac{14}{20}$$

• The fractions compare as the numerators compare.

$$\frac{12}{20} < \frac{14}{20} < \frac{15}{20}$$

So, $\frac{3}{5} < \frac{7}{10} < \frac{3}{4}$.

 Check It Out

Order the fractions from least to greatest.

8 $\frac{5}{6}, \frac{5}{7}, \frac{3}{4}, \frac{2}{3}$

9 $\frac{2}{3}, \frac{9}{10}, \frac{7}{8}, \frac{3}{4}$

10 $\frac{1}{8}, \frac{3}{4}, \frac{5}{12}, \frac{3}{8}, \frac{5}{6}$

2·2 Exercises

Compare each fraction. Use <, >, or =.

1. $\dfrac{7}{12} \,\square\, \dfrac{5}{6}$

2. $\dfrac{2}{3} \,\square\, \dfrac{5}{9}$

3. $\dfrac{3}{8} \,\square\, \dfrac{1}{3}$

4. $\dfrac{1}{6} \,\square\, \dfrac{2}{9}$

5. $\dfrac{8}{9} \,\square\, \dfrac{17}{18}$

Compare each mixed number. Use <, >, or =.

6. $3\dfrac{2}{5} \,\square\, 2\dfrac{4}{5}$

7. $1\dfrac{2}{3} \,\square\, 1\dfrac{5}{9}$

8. $2\dfrac{4}{7} \,\square\, 2\dfrac{5}{12}$

9. $4\dfrac{2}{5} \,\square\, 4\dfrac{3}{7}$

Order the fractions and mixed numbers from least to greatest.

10. $\dfrac{3}{8}, \dfrac{2}{5}, \dfrac{7}{20}$

11. $\dfrac{2}{6}, \dfrac{8}{21}, \dfrac{4}{14}$

12. $\dfrac{7}{12}, \dfrac{23}{40}, \dfrac{8}{15}, \dfrac{19}{30}$

13. $1\dfrac{8}{11}, 2\dfrac{1}{4}, 1\dfrac{3}{4}$

Use the following information to answer Exercises 14 and 15.

Baskets Made at Recess	
Toshi	$\dfrac{5}{7}$
Vanessa	$\dfrac{8}{12}$
Sylvia	$\dfrac{4}{9}$
Derrick	$\dfrac{7}{10}$

14. Who was more accurate in shots, Vanessa or Toshi?

15. Order the players from least accurate to most accurate shots.

2·3 Addition and Subtraction of Fractions

Adding and Subtracting Fractions with Like Denominators

When you add or subtract like fractions, you add or subtract the numerators and write the result over the denominator. You can also use fraction drawings to model the addition and subtraction of fractions with like denominators.

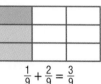

$$\frac{1}{9} + \frac{2}{9} = \frac{3}{9}$$

EXAMPLE | **Adding and Subtracting Fractions with Like Denominators**

Add $\frac{5}{12} + \frac{11}{12}$.

$5 + 11 = 16$

$\frac{5}{12} + \frac{11}{12} = \frac{16}{12}$

$\frac{16}{12} = 1\frac{4}{12} = 1\frac{1}{3}$

So, $\frac{5}{12} + \frac{11}{12} = 1\frac{1}{3}$.

- Add or subtract the numerators.
- Write the result over the denominator.
- Simplify, if possible.

Check It Out

Add or subtract. Simplify, if possible.

① $\frac{4}{5} + \frac{3}{5}$

② $\frac{11}{12} + \frac{7}{12}$

③ $\frac{8}{9} - \frac{3}{9}$

④ $\frac{15}{26} - \frac{7}{26}$

Adding and Subtracting Fractions with Unlike Denominators

You can use models to add fractions with unlike denominators.

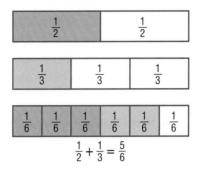

$$\frac{1}{2} + \frac{1}{3} = \frac{5}{6}$$

To add or subtract fractions with unlike denominators, you need to rename the fractions as equivalent fractions with common, or like, denominators before you find the sum or difference.

EXAMPLE | **Adding and Subtracting Fractions with Unlike Denominators**

Add $\frac{2}{3} + \frac{3}{4}$.

12 is the LCD of 3 and 4.
- Find the least common denominator of the fractions.

$\frac{2}{3} = \frac{2}{3} \times \frac{4}{4} = \frac{8}{12}$ and
- Write equivalent fractions with the LCD.

$\frac{3}{4} = \frac{3}{4} \times \frac{3}{3} = \frac{9}{12}$

$\frac{8}{12} + \frac{9}{12} = \frac{17}{12}$
- Add or subtract the numerators. Put the result over the common denominator.

$\frac{17}{12} = 1\frac{5}{12}$
- Simplify, if possible.

So, $\frac{2}{3} + \frac{3}{4} = 1\frac{5}{12}$.

 Check It Out

Add or subtract. Simplify, if possible.

5 $\frac{3}{8} + \frac{3}{4}$ **6** $\frac{5}{6} - \frac{3}{5}$ **7** $\frac{5}{9} + \frac{1}{9}$

Adding and Subtracting Mixed Numbers

Adding and subtracting mixed numbers is similar to adding and subtracting fractions.

Adding Mixed Numbers with Common Denominators

You add mixed numbers with like fractions by adding the fraction part and then the whole numbers. Sometimes you will have to simplify an improper fraction in the answer.

EXAMPLE Adding Mixed Numbers with Common Denominators

Add $7\frac{2}{5} + 6\frac{4}{5}$.

$$\left.\begin{array}{r} 7\frac{2}{5} \\ + 6\frac{4}{5} \end{array}\right\}$$

Add the whole numbers. / Add the fractions.

$$13\frac{6}{5}$$

Simplify, if possible. $13\frac{6}{5} = 14\frac{1}{5}$

So, $7\frac{2}{5} + 6\frac{4}{5} = 14\frac{1}{5}$.

 Check It Out

Add or subtract. Simplify, if possible.

8 $4\frac{2}{5} + 1\frac{3}{5}$

9 $12\frac{3}{8} - 2\frac{1}{8}$

10 $24\frac{5}{8} - 19\frac{3}{8}$

11 $17\frac{1}{4} + 2\frac{3}{4}$

Adding Mixed Numbers with Unlike Denominators

You can use fraction drawings to model the addition of mixed numbers with unlike fractions.

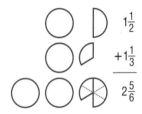

To add mixed numbers with unlike fractions, you need to write equivalent fractions with a common denominator.

EXAMPLE Adding Mixed Numbers with Unlike Denominators

Add $2\frac{2}{3} + 4\frac{1}{2}$.

$\frac{2}{3} = \frac{4}{6}$ and $\frac{1}{2} = \frac{3}{6}$

- Write the fractions with a common denominator.

Add the whole numbers.
$\left.\begin{array}{r} 2\frac{4}{6} \\ + 4\frac{3}{6} \end{array}\right\}$ Add the fractions.

- Then add and simplify.

$6\frac{7}{6}$

$6\frac{7}{6} = 7\frac{1}{6}$

- Simplify, if possible.

So, $2\frac{2}{3} + 4\frac{1}{2} = 7\frac{1}{6}$.

 Check It Out

Add. Simplify, if possible.

12 $3\frac{5}{9} + 4\frac{1}{6}$

13 $4\frac{3}{5} + 5\frac{11}{15}$

14 $4\frac{1}{9} + 3\frac{3}{4}$

15 $5\frac{1}{5} + 2\frac{3}{10}$

Subtracting Mixed Numbers

You can model the subtraction of mixed numbers with unlike fractions.

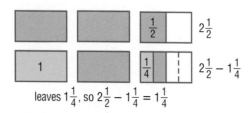

leaves $1\frac{1}{4}$, so $2\frac{1}{2} - 1\frac{1}{4} = 1\frac{1}{4}$

To subtract mixed numbers, you need to have like fractions.

EXAMPLE   Subtracting Mixed Numbers

Subtract $9\frac{1}{5} - 4\frac{2}{3}$.

- If you have unlike fractions, write them as equivalent fractions with a common denominator.

  $9\frac{1}{5} = 9\frac{3}{15}$ and $4\frac{2}{3} = 4\frac{10}{15}$

- Subtract and simplify.

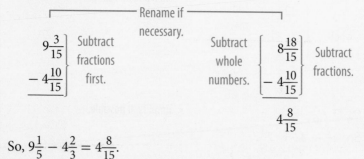

So, $9\frac{1}{5} - 4\frac{2}{3} = 4\frac{8}{15}$.

Check It Out

Subtract. Simplify, if possible.

16 $8\frac{1}{2} - 4\frac{2}{3}$

17 $7\frac{1}{10} - 4\frac{3}{5}$

18 $11\frac{3}{8} - 6\frac{1}{8}$

19 $3\frac{1}{4} - 1\frac{3}{4}$

2·3 Exercises

Add or subtract.

1. $\frac{5}{7} + \frac{6}{11}$

2. $\frac{1}{4} + \frac{1}{3}$

3. $\frac{2}{5} - \frac{1}{10}$

4. $\frac{7}{8} - \frac{3}{4}$

5. $\frac{7}{12} - \frac{3}{10}$

6. $\frac{3}{10} + \frac{4}{5}$

7. $5\frac{2}{3} + 2\frac{1}{2}$

8. $4\frac{1}{2} - 3\frac{3}{4}$

9. $1\frac{2}{3} + 1\frac{1}{4}$

10. $7\frac{3}{5} - 3\frac{2}{3}$

11. $2\frac{3}{4} + 6\frac{5}{16}$

12. $9\frac{4}{7} - 5\frac{1}{14}$

13. $8\frac{3}{8} + 6\frac{3}{4}$

14. $10\frac{1}{10} - 3\frac{3}{20}$

15. $9\frac{1}{2} - 4\frac{7}{8}$

16. $2\frac{1}{3} - 1\frac{2}{3}$

17. Desrie is planning to make a two-piece costume. One piece requires $1\frac{5}{8}$ yd of material, and the other requires $1\frac{3}{4}$ yd. She has $4\frac{1}{2}$ yd of material. Does she have enough to make the costume?

18. Atiba was $53\frac{7}{8}$ in. tall on his birthday last year. On his birthday this year, he was $56\frac{1}{4}$ in. tall. How much did he grow during the year?

19. Hadas's punch bowl holds 8 qt. Can she serve cranberry punch made with $6\frac{2}{3}$ qt cranberry juice and $2\frac{1}{4}$ qt apple juice?

20. Nia jogged $4\frac{1}{10}$ mi on Sunday, $2\frac{2}{5}$ mi on Tuesday, and $3\frac{1}{2}$ mi on Thursday. How many miles does she have to jog on Saturday to reach her weekly goal of $12\frac{1}{2}$ mi?

2•4 Multiplication and Division of Fractions

Multiplying Fractions

You know that 3×2 means "3 groups of 2." Multiplying fractions involves the same concept: $3 \times \frac{1}{4}$ means "3 groups of $\frac{1}{4}$." You may find it helpful to know that, in math, the word *times* frequently means to multiply.

1 group of $\frac{1}{4}$ ➡

3 groups of $\frac{1}{4}$ or $\frac{3}{4}$

The same is true when you are multiplying a fraction by a fraction. For example, $\frac{1}{4} \times \frac{1}{2}$ means that you actually find $\frac{1}{4}$ of $\frac{1}{2}$.

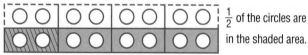

$\frac{1}{2}$ of the circles are in the shaded area.

$\frac{1}{4}$ of $\frac{1}{2}$ of the circles are $\frac{1}{8}$ of *all* the circles.   So, $\frac{1}{4} \times \frac{1}{2} = \frac{1}{8}$.

When you are not using models to multiply fractions, you multiply the numerators and then the denominators. There is no need to find a common denominator.

EXAMPLE **Multiplying Fractions**

Multiply $\frac{2}{3} \times 2\frac{1}{5}$.

$\frac{2}{3} \times 2\frac{1}{5} = \frac{2}{3} \times \frac{11}{5}$

$\frac{2}{3} \times \frac{11}{5} = \frac{2 \times 11}{3 \times 5} = \frac{22}{15}$

$\frac{22}{15} = 1\frac{7}{15}$

So, $\frac{2}{3} \times 2\frac{1}{5} = 1\frac{7}{15}$.

• Convert mixed numbers, if any, to *improper fractions* (p. 101).
• Multiply the numerators and the denominators.
• Write the products in simplest form.

Check It Out

Multiply.

1 $\frac{1}{2} \times \frac{3}{5}$

2 $3\frac{1}{3} \times \frac{1}{2}$

Shortcut for Multiplying Fractions

You can use a shortcut when you multiply fractions. Instead of multiplying across and then writing the product in simplest form, you can simplify the **factors** first.

EXAMPLE	Multiplying Factors

Multiply $\frac{4}{5} \times \frac{15}{16}$.

$\frac{4}{5} \times \frac{15}{16}$
- Write mixed numbers, if any, as improper fractions.

$\frac{4}{5} \times \frac{5 \times 3}{4 \times 4}$
- Write as factors, if possible.

$\frac{4 \times 5 \times 3}{4 \times 5 \times 4}$
- Combine factors on top and bottom. You can reorder by the Communicative Property of Multiplication.

$= 1 \times 1 \times \frac{3}{4}$
- Divide the common factors by 1.

$= \frac{3}{4}$
- Write the product.

You can use a shorthand method to show the steps above.

$\frac{\overset{1}{4}}{\underset{1}{5}} \times \frac{\overset{1}{5} \times 3}{\underset{1}{4} \times 4}$
- Denote the factors that divide to 1 by marking them as shown.

If you do not find all the common factors by using this method, you will need to write your answers in simplest form.

Check It Out

Multiply. Simplify, if possible.

3 $\frac{4}{7} \times \frac{21}{24}$

4 $\frac{3}{5} \times \frac{20}{21}$

5 $3\frac{1}{5} \times 1\frac{1}{4}$

Finding the Reciprocal of a Number

If the product of two numbers is 1, the numbers are called multiplicative inverses, or **reciprocals**.

$$\frac{3}{8} \times \frac{8}{3} = \frac{24}{24} = 1$$

$\frac{8}{3}$ is the reciprocal of $\frac{3}{8}$ because $\frac{3}{8} \times \frac{8}{3} = 1$.

To find the reciprocal of a number, you switch the numerator and the denominator. When you use a reciprocal, it is the same as using an inverse operation.

Number	$\frac{3}{5}$	$2 = \frac{2}{1}$	$3\frac{1}{2} = \frac{7}{2}$
Reciprocal	$\frac{5}{3}$	$\frac{1}{2}$	$\frac{2}{7}$

The number 0 does not have a reciprocal.

Check It Out

Find the reciprocal of each number.

6 $\frac{2}{5}$

7 4

8 $2\frac{1}{3}$

9 $3\frac{4}{7}$

Dividing Fractions

When you divide a fraction by a fraction, such as $\frac{1}{2} \div \frac{1}{6}$, you are really finding out how many $\frac{1}{6}$ are in $\frac{1}{2}$. That's why the answer is 3. To divide fractions, you replace the divisor with its reciprocal and then multiply to get your answer.

$$\frac{1}{2} \div \frac{1}{6} = \frac{1}{2} \times \frac{6}{1} = 3$$

EXAMPLE Dividing Fractions

Divide $\frac{4}{5} \div 2\frac{2}{3}$.

$\frac{4}{5} \div \frac{8}{3}$
- Write any mixed numbers as improper fractions.

$\frac{4}{5} \times \frac{3}{8} = \frac{\overset{1}{\cancel{4}}}{5} \times \frac{3}{\underset{2}{\cancel{8}}} = \frac{1}{5} \times \frac{3}{2}$
- Replace the divisor with its reciprocal and simplify factors.

$\frac{1}{5} \times \frac{3}{2} = \frac{3}{10}$
- Multiply.

So, $\frac{4}{5} \div 2\frac{2}{3} = \frac{3}{10}$.

Check It Out

Divide.

10 $\frac{2}{3} \div \frac{1}{8}$

11 $\frac{1}{4} \div \frac{1}{14}$

12 $\frac{3}{10} \div \frac{2}{5}$

13 $3\frac{1}{2} \div \frac{2}{5}$

2•4 Exercises

Multiply.

1. $\frac{2}{3} \times \frac{7}{8}$

2. $\frac{1}{8} \times 2\frac{1}{2}$

3. $3\frac{1}{3} \times \frac{3}{5}$

4. $\frac{4}{9} \times 1\frac{1}{2}$

5. $\frac{6}{8} \times 3\frac{1}{4}$

6. $2\frac{1}{2} \times 1\frac{1}{3}$

7. $\frac{7}{8} \times 1\frac{2}{7}$

8. $1\frac{1}{3} \times 1\frac{1}{2}$

9. $3\frac{7}{8} \times \frac{9}{12}$

10. $\frac{3}{4} \times \frac{6}{11}$

11. $6\frac{3}{9} \times 2\frac{8}{14}$

12. $\frac{4}{5} \times \frac{6}{7}$

Find the reciprocal of each number.

13. $\frac{4}{5}$

14. 3

15. $4\frac{1}{2}$

16. $4\frac{2}{3}$

17. $\frac{6}{7}$

Divide.

18. $\frac{2}{3} \div \frac{1}{2}$

19. $\frac{5}{8} \div \frac{3}{4}$

20. $1\frac{3}{4} \div \frac{5}{6}$

21. $2\frac{1}{3} \div \frac{2}{3}$

22. $\frac{7}{15} \div \frac{1}{6}$

23. $1\frac{1}{7} \div \frac{2}{5}$

24. $\frac{3}{16} \div \frac{1}{4}$

25. $2\frac{1}{3} \div 9$

26. $5\frac{1}{5} \div \frac{1}{2}$

27. $3\frac{1}{2} \div 2\frac{1}{3}$

28. $2\frac{9}{24} \div 3\frac{8}{12}$

29. Naoko's spaghetti sauce recipe calls for $1\frac{1}{3}$ tsp of salt for every quart of sauce. How many teaspoons of salt will she need to make $5\frac{1}{2}$ qt of sauce?

30. Aisha plans to buy corduroy material at $4 per yd. If she needs to buy $8\frac{1}{2}$ yd of material, how much will it cost?

2·5 Naming and Ordering Decimals

Naming Decimals Greater Than and Less Than One

Decimal numbers are based on units of 10.

Place-Value Chart

					$\frac{1}{10}$	$\frac{1}{100}$	$\frac{1}{1000}$	$\frac{1}{10,000}$	$\frac{1}{100,000}$
10,000	1,000	100	10	1	0.1	0.01	0.001	0.0001	0.00001
ten-thousands	thousands	hundreds	tens	ones	tenths	hundredths	thousandths	ten-thousandths	hundred-thousandths

The chart shows the place value for some of the digits of a decimal. You can use a place-value chart to help you name decimals greater than and less than one.

EXAMPLE Naming Decimal Numbers

Read 27.6314.

- Values to the left of the decimal point are greater than one.

  27 means 2 tens and 7 ones.

- Read the decimal. The word name of the decimal is determined by the value of the digit in the last place.

  The last digit (4) is in the ten-thousandths place.

27.6314 is read as twenty-seven *and* six thousand three hundred fourteen ten-thousandths.

Check It Out

Use the place-value chart to tell what each boldfaced digit means. Then write the numbers in words.

1 4.4**1**1

2 0.0**3**2

3 5.004**6**

4 0.003**4**1

Comparing Decimals

When zeros are added to the right of the decimal in the following manner, they do not change the value of the number.

$1.039 = 1.0390 = 1.03900 = 1.039000 \ldots$

To compare decimals, compare their place values.

EXAMPLE Comparing Decimals

Compare 19.5032 and 19.5047.

• Start at the left. Find the first place where the numbers are different.

19.5032
19.5047
↑
The thousandths place is different.

• Compare the digits that are different.

3 < 4

• The numbers compare the same way the digits compare.

Thus, 19.5032 < 19.5047.

 Check It Out

Compare. Write >, <, or =.

5 26.3 □ 26.4

6 0.0176 □ 0.0071

Ordering Decimals

To write decimals from least to greatest and vice versa, you need to first compare the numbers two at a time.

Order the decimals 1.143, 0.143, and 1.10 from least to greatest.

• Compare the numbers two at a time.

  1.143 > 0.143   1.143 > 1.10   0.143 < 1.10

• List the decimals from least to greatest.

  0.143; 1.10; 1.143

Check It Out

Write in order from least to greatest.

7 3.0186, 30.618, 3.1608

8 9.083, 9.381, 93.8, 9.084, 9

9 0.622, 0.662, 0.6212, 0.6612

Rounding Decimals

Rounding decimals is similar to rounding whole numbers.

Round 14.046 to the nearest hundredth.

• Find the rounding place.

  14.046
     ↑
  hundredths

• Look at the digit to the right of the rounding place.

  14.046

If it is less than 5, leave the digit in the rounding place unchanged. If it is more than or equal to 5, increase the digit in the rounding place by 1.

  6 > 5

• Write the rounded number.

  14.046 rounded to the nearest hundredth is 14.05.

Check It Out

Round each decimal to the nearest hundredth.

10 1.544   **11** 36.389   **12** 8.299

2·5 Exercises

Tell the hundredths digit in each decimal.

1. 24.012 **2.** 6.204 **3.** 231.679 **4.** 3.298

Write what each digit means.

5. 0.165

6. 0.208

7. 0.149

Tell what each boldfaced digit means.

8. 34.2**4**1 **9.** 4.3**4**61

10. 0.129**6** **11.** 24.1**4**

Compare. Use <, >, or =.

12. 15.099 ☐ 15.11 **13.** 12.5640 ☐ 12.56

14. 7.1498 ☐ 7.2 **15.** 0.684 ☐ 0.694

List in order from least to greatest.

16. 0.909, 0.090, 0.90, 0.999

17. 8.822, 8.288, 8.282, 8.812

18. 6.85, 0.68, 0.685, 68.5

Round each decimal to the indicated place value.

19. 1.6432; thousandths

20. 48.098; hundredths

21. 3.86739; ten-thousandths

22. Four ice skaters are in a competition in which the highest possible score is 10.0. Three of the skaters have completed their performances, and their scores are 9.61, 9.65, and 9.60. What score, rounded to the nearest hundredth, must the last skater get in order to win the competition?

23. The Morales family traveled 45.66 mi on Saturday, 45.06 mi on Sunday, and 45.65 mi on Monday. On which day did they travel farthest?

2·6 Decimal Operations

Adding and Subtracting Decimals

Adding and subtracting decimals is similar to adding whole numbers. You need to be careful to line up the digits by their place.

EXAMPLE Adding and Subtracting Decimals

Add 4.75 + 0.6 + 32.46.

$$\begin{array}{r} 4.75 \\ 0.6 \\ + 32.46 \end{array}$$ • Line up the decimal points.

$$\begin{array}{r} \overset{1}{4.75} \\ 0.6 \\ + 32.46 \\ \hline 1 \end{array}$$ • Add (or subtract) the place farthest right. Regroup as necessary.

$$\begin{array}{r} \overset{1\ 1}{4.75} \\ 0.6 \\ + 32.46 \\ \hline 81 \end{array}$$ • Add (or subtract) the next place left. Regroup as necessary.

$$\begin{array}{r} 4.75 \\ 0.6 \\ + 32.46 \\ \hline 37.81 \end{array}$$ • Continue through the whole numbers. Place the decimal point in the result.

 Check It Out

Solve.

1 8.1 + 31.75

2 19.58 + 37.42 + 25.75

3 17.8 − 4.69

4 52.7 − 0.07219

Estimating Decimal Sums and Differences

One way that you can **estimate** decimal sums and differences is to use *compatible numbers*. **Compatible numbers** are close to the numbers in the problem and are easy to work with mentally.

Estimate the sum of 2.244 + 6.711.

• Replace the numbers with compatible numbers.

$$2.244 \longrightarrow 2$$
$$6.711 \longrightarrow 7$$

• Add the numbers.

$$2 + 7 = 9$$

Estimate the difference of 12.6 − 8.4.

$$12.6 \longrightarrow 13$$
$$8.4 \longrightarrow 8$$
$$13 - 8 = 5$$

Check It Out

Estimate each sum or difference.

5 8.64 + 5.33

6 11.3 − 9.4

7 18.145 − 3.66

8 3.48 + 5.14 + 8.53

Multiplying Decimals

Multiplying decimals is much the same as multiplying whole numbers. You can model the multiplication of decimals with a 10 × 10 grid. Each tiny square is equal to one hundredth.

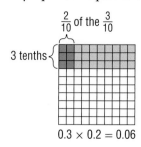

$\frac{2}{10}$ of the $\frac{3}{10}$

3 tenths

$$0.3 \times 0.2 = 0.06$$

EXAMPLE Multiplying Decimals

Multiply 32.6 × 0.08.

- Multiply as with whole numbers.

$$\begin{array}{r} 32.6 \\ \times\ 0.08 \end{array} \qquad \begin{array}{r} 326 \\ \times\ \ 8 \\ \hline 2608 \end{array}$$

- Add the number of decimal places for the factors.

32.6 ⟶ 1 decimal place
0.08 ⟶ 2 decimal places
1 + 2 = 3 total decimal places

- Place the decimal point in the product.

$$\begin{array}{r} 32.6 \\ \times\ 0.08 \\ \hline 2.608 \end{array}$$ → 1 decimal place
→ 2 decimal places
→ 3 decimal places

So, 32.6 × 0.08 = 2.608.

 Check It Out

Multiply.

⑨ 10.9 × 0.7
⑩ 6.2 × 0.087
⑪ 0.61 × 3.2
⑫ 9.81 × 6.5

Multiplying Decimals with Zeros in the Product

Sometimes when you are multiplying decimals, you need to add zeros in the product.

Multiply 0.7 × 0.0345.

- Multiply as with whole numbers.

$$
\begin{array}{r}
0.0345 \\
\times\ 0.7 \\
\hline
\end{array}
\qquad
\begin{array}{r}
00345 \\
\times\ 07 \\
\hline
2415
\end{array}
$$

- Count the number of decimal places in the factors.

$$
\begin{array}{r}
00345 \\
\times\ 07 \\
\hline
2415
\end{array}
\begin{array}{l}
\rightarrow\ 4\ \text{decimal places} \\
\rightarrow\ 1\ \text{decimal place} \\
\rightarrow\ \text{needs 5 decimal places}
\end{array}
$$

- Add zeros in the product as necessary.

.02415

Because 5 places are needed in the product, write a zero to the left of the 2.

0.7 × 0.0345 = 0.02415

 Check It Out

Multiply.

13 4.1 × 0.0037

14 0.961 × 0.05

Estimating Decimal Products

To estimate decimal products, you replace given numbers with compatible numbers. Compatible numbers are estimates you choose because they are easier to work with mentally.

To estimate the product of 27.3 × 44.2, you start by replacing the factors with compatible numbers. 27.3 becomes 30. 44.2 becomes 40. Then multiply mentally. 30 × 40 = 1,200. So, 27.3 × 44.2 is about 1,200.

Check It Out

Estimate each product using compatible numbers.

15 25.71 × 9.4

16 9.48 × 10.73

DECIMAL OPERATIONS **2·6**

Dividing Decimals

Dividing decimals is similar to dividing whole numbers.
You can use a model to help you understand dividing decimals.
For example, $0.9 \div 0.3$ means how many groups of 0.3 are in 0.9?

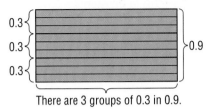

There are 3 groups of 0.3 in 0.9.

There are 3 groups of 0.3 in 0.9, so $0.9 \div 0.3 = 3$.

EXAMPLE Dividing Decimals

Divide $31.79 \div 1.1$.

$1.1\overline{)31.79}$

$1.1 \times 10 = 11$
- Multiply the divisor by a power of ten to make it a whole number.

$31.79 \times 10 = 317.9$
$11.\overline{)317.9}$
- Multiply the dividend by the same power of ten.

$11.\overline{)317.9}^{\,\cdot}$
- Place the decimal point in the quotient.

$11.\overline{)317.9}^{\,28.9}$
- Divide.

So, $31.79 \div 1.1 = 28.9$.

 Check It Out

Divide.

17 $0.231 \div 0.07$ **18** $0.312 \div 0.06$

19 $1.22 \div 0.4$ **20** $0.6497 \div 8.9$

Zeros in Division

You can use zeros to hold places in the dividend when you are dividing decimals.

EXAMPLE **Using Zeros in Division**

Divide $879.9 \div 4.2$.

$4.2\overline{)879.9}$

$4.2 \times 10 = 42$

$879.9 \times 10 = 8799$

- Multiply the divisor and the dividend by a power of ten. Place the decimal point.

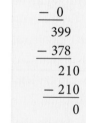

$$\begin{array}{r} 209. \\ 42.\overline{)8799.} \\ -84 \\ \hline 39 \\ -0 \\ \hline 399 \\ -378 \\ \hline 21 \end{array}$$

- Divide.

$$\begin{array}{r} 209.5 \\ 42.\overline{)8799.0} \\ -84 \\ \hline 39 \\ -0 \\ \hline 399 \\ -378 \\ \hline 210 \\ -210 \\ \hline 0 \end{array}$$

- Use zeros to hold places in the dividend. Continue to divide until the remainder is zero.

So, $879.9 \div 4.2 = 209.5$.

Check It Out

Divide until the remainder is zero.

21. $298.116 \div 7.35$
22. $89.61 \div 2.9$

Rounding Decimal Quotients

You can use a calculator to divide decimals and round quotients.

Divide 6.3 ÷ 2.6. Round the quotient to the nearest hundredth.

• Use your calculator to divide.

  6.3 ÷ 2.6 = 2.4230769

• To round the quotient, look at one place to the right of the
  rounding place. If the digit to the right of the rounding place is
  5 or above, round up. If the digit to the right of the rounding
  place is less than 5, the digit to be rounded remains the same.

  2.4230769   3 < 5, so 2.4230769 rounded to the nearest
  hundredth is 2.42.

6.3 ÷ 2.6 = 2.42

 Check It Out

Use a calculator to find each quotient. Round to the nearest
hundredth.

23 2.2 ÷ 0.3

24 32.5 ÷ 0.32

25 0.671 ÷ 2.33

APPLICATION **Luxuries or Necessities?**

Recent economic reforms have made China one of the fastest-growing economies in the world. After years of hardship, the country has not yet caught up in providing luxuries for its vast population of approximately 1,200,000,000.

Standard of living refers to the level of goods, services, and luxuries available to an individual or a population. Here are two examples of how China's standard of living compares with that of the United States.

	China	United States
Number of people per telephone	36.4	1.3
Number of people per TV	6.7	1.2

When China has the same number of telephones per person as the United States has, how many will it have? See **Hot**Solutions for answer.

2·6 Exercises

Estimate each sum or difference.

   1. 3.24 + 1.06   **2.** 6.09 − 3.7

   3. 8.445 + 0.92   **4.** 3.972 + 4.124

   5. 11.92 − 8.0014

Add.

   6. 234.1 + 4.92   **7.** 65.11 + 22.64

   8. $11.19 + $228.16   **9.** 7.0325 + 0.81

  10. 1.8 + 4 + 2.6473

Subtract.

  11. 22 − 11.788   **12.** 42.108 − 0.843

  13. 386.1 − 2.94   **14.** 52.12 − 18.666

  15. 12.65 − 3.0045

Multiply.

  16. 0.7 × 6.633   **17.** 12.6 × 33.44

  18. 0.14 × 0.02   **19.** 49.32 × 0.6484

  20. 0.57 × 0.91

Divide.

  21. 43.68 ÷ 5.2   **22.** 6.552 ÷ 9.1

  23. 65.026 ÷ 0.82   **24.** 2.175 ÷ 2.9

Divide. Round the quotient to the nearest hundredth.

  25. 18.47 ÷ 5.96   **26.** 18.6 ÷ 2.8

  27. 82.3 ÷ 8.76   **28.** 63.7 ÷ 7.6

  29. Bananas at the Quick Stop Market cost $0.49 per lb. Find the cost of 4.6 lb of bananas. Round your answer to the nearest cent.

  30. How much change would you get from a $50 bill if you bought two $14.95 shirts?

2·7 Meaning of Percent

Naming Percents

A **ratio** of a number to 100 is called a **percent**. Percent means *per hundred* and is represented by the symbol %.

You can use graph paper to model percents. There are 100 squares on a 10-by-10 sheet of graph paper. So, a 10-by-10 sheet can be used to represent 100%. Because percent means how many out of 100, it is easy to tell what percent of the 100-square graph paper is shaded.

25 of 100 are blue (25% blue). 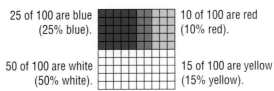 10 of 100 are red (10% red).

50 of 100 are white (50% white). 15 of 100 are yellow (15% yellow).

Check It Out

What percent of each square is shaded?

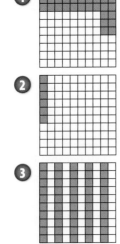

1

2

3

Understanding the Meaning of Percent

One way to think about percents is to become comfortable with a few **benchmarks**. You build what you know about percents on these different benchmarks. You can use these benchmarks to help you estimate percents of other things.

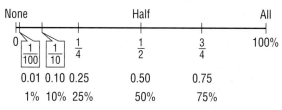

EXAMPLE **Estimating Percents**

Estimate 48% of 50.

- Choose a benchmark, or combination of benchmarks, close to the target percent.

    48% is close to 50%.

- Find the fraction or decimal equivalent to the benchmark percent.

    $50\% = \frac{1}{2}$

- Use the benchmark equivalent to estimate the percent.

    $\frac{1}{2}$ of 50 is 25.

So, 48% of 50 is about 25.

Check It Out

Use fractional benchmarks to estimate the percents.

①② **4** 49% of 100

5 24% of 100

6 76% of 100

7 11% of 100

Using Mental Math to Estimate Percent

You can use fraction or decimal benchmarks to help you estimate the percent of something, such as a tip in a restaurant.

Estimate a 10% tip for a bill of $4.48.

- Round to a convenient number.
  $4.48 rounds to $4.50.
- Think of the percent as a benchmark or combination of benchmarks.
  $10\% = 0.10 = \frac{1}{10}$
- Multiply mentally. Combine, if necessary.
  $0.10 \times \$4.50 = \$0.45$ The tip is about $0.45.

EXAMPLE   Estimating a Tip

Estimate a 20% tip for a bill of $38.70.

$38.70 rounds to $39.00	• Round to a convenient number.
$20\% = 0.20$	• Think of the percent as a benchmark.
$0.20 \times \$39.00 = \$7.80$	• Multiply mentally.

The tip is about $8.00.

 Check It Out

Estimate the amount of each tip.

8 10% of $2.50

9 20% of $85

2·7 Exercises

Write the percent that is shaded.

1.

2.

3.

4.

5.

6.

Use fractional benchmarks to estimate the percent of each number.

7. 26% of 80

8. 45% of 200

9. 12% of 38

10. 81% of 120

11. Estimate a 15% tip for a $60 bill.

12. Estimate a 20% tip for a $39 bill.

13. Estimate a 10% tip for an $84 bill.

14. Estimate a 25% tip for a $75 bill.

15. Estimate a 6% tip for a $54 bill.

2·8 Using and Finding Percents

Finding a Percent of a Number

You can use decimals or fractions to find the percent of a number. You first change the percent to a decimal or a fraction. Sometimes it is easier to change to a decimal representation and other times to a fractional one.

You can use either the fraction method or the decimal method to find 25% of 60.

EXAMPLE Finding the Percent of a Number: Two Methods

Find 25% of 60.

Decimal Method
- Change percent to decimal.

  $25\% = 0.25$

- Multiply.

  $0.25 \times 60 = 15$

So, 25% of 60 = 15.

Fraction Method
- Change percent to fraction.

  $25\% = \dfrac{25}{100}$

- Reduce fraction to lowest terms.

  $\dfrac{25}{100} = \dfrac{1}{4}$

- Multiply.

  $\dfrac{1}{4} \times \dfrac{60}{1} = \dfrac{60}{4} = 15$

So, 25% of 60 = 15.

Check It Out

Find the percent of each number.

1. 65% of 80
2. 85% of 500
3. 30% of 90
4. 75% of 420

Finding the Percent of a Number: Proportion Method

You can also use proportions to help you find the percent of a number.

EXAMPLE Finding the Percent of a Number: Proportion Method

Gwen works in a music store. She receives a commission of 10% on her sales. Last month she sold $850 worth of CDs. What was her commission?

- Use a proportion (p. 261) to find the percent of a number.

  P = Part (of the whole or total) $\dfrac{P}{R} = \dfrac{W}{100}$

  W = Whole (total)

  R = Rate (percentage)

- Identify the given items before trying to find the unknown.

  P is unknown; call it x.

  R is 10%.

  W is $850.

- Set up the proportion.

  $\dfrac{P}{R} = \dfrac{W}{100}$ $\dfrac{x}{10} = \dfrac{850}{100}$

- Cross multiply.

  $100x = 8{,}500$

- Divide both sides of the equation by 100.

  $\dfrac{8{,}500}{100} = \dfrac{100x}{100}$

  $85 = x$

Gwen received a commission of $85.

Check It Out

Use a proportion to find the percent of each number.

5 76% of 39

6 14% of 85

7 66% of 122

8 55% of 300

Finding Percent and Whole

Setting up and solving a proportion can help you find what percent a number is of a second number. Use the ratio $\frac{P}{W} = \frac{R}{100}$ where P = Part (of whole), W = Whole, and R = Rate (percentage).

EXAMPLE Finding the Percent

What percent of 40 is 15?

- Set up a proportion using this form.

$$\frac{\text{Part}}{\text{Whole}} = \frac{\text{Percent}}{100}$$

$$\frac{15}{40} = \frac{n}{100}$$

(The number after the word *of* indicates the whole.)

- Find the cross products of the proportion.

$$100 \times 15 = 40 \times n$$

- Find the products.

$$1{,}500 = 40n$$

- Divide both sides of the equation by 100.

$$\frac{1{,}500}{40} = \frac{40n}{40} \qquad \frac{1{,}500}{40} = n \qquad 37.5\% = n$$

So, 15 is 37.5% of 40.

 Check It Out

Solve.

9 What percent of 240 is 80?

10 What percent of 64 is 288?

11 What percent of 2 is 8?

12 What percent of 55 is 33?

8 is 32% of what number?

- Set up a percent proportion using this form.

$$\frac{\text{Part}}{\text{Whole}} = \frac{\text{Percent}}{100}$$

$$\frac{8}{n} = \frac{32}{100}$$

(The phrase *what number* after the word *of* is the whole.)

- Find the cross products of the proportion.

$$8 \times 100 = 32 \times n$$

- Find the products.

$$800 = 32n$$

- Divide both sides of the equation.

$$\frac{800}{32} = \frac{32n}{32}$$

$$n = 25$$

So, 8 is 32% of 25.

Check It Out

Find the whole.

⑬ 52 is 50% of what number?

⑭ 15 is 75% of what number?

⑮ 40 is 160% of what number?

⑯ 84 is 7% of what number?

Estimating a Percent of a Number

To estimate a percent of a number, you can use what you know about *compatible numbers* and simple fractions.

The table can help you estimate the percent of a number.

Percent	1%	5%	10%	20%	25%	$33\frac{1}{3}$%	50%	$66\frac{2}{3}$%	75%	100%
Fraction	$\frac{1}{100}$	$\frac{1}{20}$	$\frac{1}{10}$	$\frac{1}{5}$	$\frac{1}{4}$	$\frac{1}{3}$	$\frac{1}{2}$	$\frac{2}{3}$	$\frac{3}{4}$	1

EXAMPLE Estimating a Percent of a Number

Estimate 12% of 32.

- Find the percent that is closest to the percent you want to find.

    12% is about 10%.

- Find the fractional equivalent for the percent.

    10% is equivalent to $\frac{1}{10}$.

- Find a compatible number for the number you want to find the percent of.

    32 is about 30.

- Use the fraction to find the percent.

    $\frac{1}{10}$ of 30 is 3.

So, 12% of 32 is about 3.

 Check It Out

Estimate the percent.

17 19% of 112

18 65% of 298

Percent of Increase or Decrease

Sometimes it is helpful to keep a record of your monthly expenses. Keeping a record allows you to see the actual *percent of increase* or *decrease* in your expenses. You can make a chart to record your expenses.

Expenses	November	December	Amount of Increase or Decrease	% of Increase or Decrease
Clothing	$ 125	$ 76	$49	39%
Entertainment	$ 44	$ 85		
Food	$ 210	$199	$ 11	5%
Miscellaneous	$ 110	$ 95		
Rent	$ 450	$465	$ 15	3%
Travel	$ 83	$ 65		
Total	$1,022	$985	$37	4%

EXAMPLE Finding the Percent of Increase

According to the table, how much was the percent of increase for entertainment?

$85 - 44 = 41$ • Subtract the original amount from the new amount to find the amount of increase.

$$
\begin{array}{r}
0.931818 \\
44\overline{)41.000000} \\
-\,396 \\
\hline
140 \\
-\,132 \\
\hline
80 \\
-\,44 \\
\hline
360 \\
-\,352 \\
\hline
80 \\
-\,44 \\
\hline
360 \\
-\,352 \\
\hline
8
\end{array}
$$

• Divide the amount of increase by the original amount.

$0.931818 = 0.93 = 93\%$ • Round to the nearest hundredth, and convert to a percent.

The percent of increase from $44 to $85 is 93%.

EXAMPLE **Finding the Percent of Decrease**

During November, $83 was spent on travel. In December, $65 was spent on travel. How much was the percent of decrease?

$83 - 65 = 18$

- Subtract the new amount from the original amount to find the amount of decrease.

$$
\begin{array}{r}
0.216867 \\
83\overline{)18.000000} \\
-166 \\
\hline
140 \\
-83 \\
\hline
570 \\
-498 \\
\hline
720 \\
-664 \\
\hline
560 \\
-498 \\
\hline
620 \\
-581 \\
\hline
39
\end{array}
$$

- Divide the amount of decrease by the original amount.

$0.216867 = 0.22 = 22\%$

- Round to the nearest hundredth, and convert to a percent.

The percent of decrease from $83 to $65 is 22%.

Check It Out

Use a calculator to find the percent of increase or decrease.

19 16 to 38

20 43 to 5

21 99 to 3

22 6 to 10

Discounts and Sale Prices

A **discount** is the amount that an item is reduced from the regular price. The sale price is the regular price minus the discount. Discount stores have regular prices below the suggested retail price. You can use percents to find discount and resulting sale prices.

This CD player has a regular price of $109.99. It is on sale for 25% off the regular price. How much money will you save by buying the item on sale?

EXAMPLE Finding Discounts and Sale Prices

The regular price of an item is $109.99. It is marked 25% off. Find the discount and the sale price.

$$
\begin{array}{r}
\$109.99 \\
\times\ 0.25 \\
\hline
54995 \\
+\ 21998 \\
\hline
27.4975
\end{array}
$$

• Multiply the regular price by the discount.

$27.4975 = 27.50$

• If necessary, round the discount to the nearest hundredth.

The discount is $27.50.

$$
\begin{array}{r}
\$109.99 \\
-\ 27.50 \\
\hline
\$\ 82.49
\end{array}
$$

• Subtract the discount from the regular price. This will give you the sale price.

The sale price is $82.49.

Check It Out

Find the discount and sale price.

㉓ regular price: $90, discount percent: 40%

㉔ regular price: $120, discount percent: 15%

Sales Tax

You can use the tax rate to help you find the amount of sales tax on a purchase. Add the amount of the tax to the purchase price to find the total cost.

Estimate a 7% sales tax on a purchase of $27.36.

• Think of the percent as a benchmark.

$$7\% = 0.07 = \frac{7}{100}$$

• Multiply.

$$27.36 \times 0.07 = 1.915$$

Round to the nearest hundredth: $1.92.

• Add the tax to the purchase to find the total.

$$\$27.36 + \$1.92 = \$29.28$$

Check It Out

Find the final cost of each purchase.

㉕ 8% tax on $84.34

㉖ 20% tax on $1,289.78

㉗ 7% tax on $167.13

Finding Simple Interest

When you have a savings account, the bank pays you for the use of your money. When you take out a loan from a bank, you pay the bank for the use of their money. In both situations, the money paid is called the *interest*. The amount of money you borrow or have in your savings account is called the *principal*. To find out how much interest you will pay or earn, you can use the formula $I = p \times r \times t$. The table below can help you understand the formula.

P	Principal—the amount of money you borrow or save
R	Interest Rate—the percent of the principal you pay or earn
T	Time—the length of time you borrow or save (in years)
I	Total Interest—the interest you pay or earn for the entire time
A	Amount—total amount (principal plus interest) you pay or earn

EXAMPLE Finding Simple Interest

Find the total amount you will pay if you borrow $5,000 at 7% simple interest for 3 years.

$p \times r \times t = I$

5,000	350
× 0.07	× 3
350	1050

$1,050 is the interest.

- Use the formula to find the total amount you will pay.

$p + I = $ total amount

$5,000	
+ 1,050	
$6,050	

- To find the total amount you will pay back, add the principal and the interest.

$6,050 is the total amount of money to be paid back.

Check It Out
Find the interest (*I*) and the total amount.

28 $P = \$750$, $R = 13\%$, $T = 2$ years

29 $P = \$3,600$, $R = 14\%$, $T = 9$ months

2•8 Exercises

Find the percent of each number.

1. 3% of 45

2. 44% of 125

3. 95% of 64

4. 2% of 15.4

Solve.

5. What percent of 40 is 29?

6. 15 is what percent of 60?

7. 4 is what percent of 18?

8. What percent of 5 is 3?

9. 64 is what percent of 120?

Solve.

10. 40% of what number is 30?

11. 25% of what number is 11?

12. 96% of what number is 24?

13. 67% of what number is 26.8?

14. 62% of what number is 15.5?

Find the percent of increase or decrease to the nearest percent.

15. 8 to 10

16. 45 to 18

17. 12 to 4

18. 15 to 20

Find the discount and sale price.

19. regular price: $19.95, discount percent: 40%

20. regular price: $65.99, discount percent: 15%

21. regular price: $285.75, discount percent: 22%

22. regular price: $385, discount percent: 40%

Find the final cost of each purchase.

23. 20% tax on $1,608.98

24. 6% tax on $278.13

Find the interest and total amount.

25. $P = \$8,500$, $R = 6.5\%$ per year, $T = 1$ year

26. $P = \$1,200$, $R = 7\%$ per year, $T = 2$ years

27. $P = \$2,400$, $R = 11\%$ per year, $T = 6$ months

Estimate the percent of each number.

28. 15% of 65

29. 27% of 74

30. 76% of 124

31. Lyudmila bought a CD player for 25% off the regular price of $179.99. How much did she save? How much did she pay?

32. A DVD player is on sale for 15% off the regular price of $289.89. What is the discount? What is the sale price?

2·9 Fraction, Decimal, and Percent Relationships

Percents and Fractions

Percents describe a ratio out of 100. A percent can be written as a fraction with a denominator of 100. The table shows how some percents are written as fractions.

Percent	Fraction
50 out of 100 = 50%	$\frac{50}{100} = \frac{1}{2}$
$33\frac{1}{3}$ out of 100 = $33\frac{1}{3}$%	$\frac{33.\overline{3}}{100} = \frac{1}{3}$
25 out of 100 = 25%	$\frac{25}{100} = \frac{1}{4}$
20 out of 100 = 20%	$\frac{20}{100} = \frac{1}{5}$
10 out of 100 = 10%	$\frac{10}{100} = \frac{1}{10}$
1 out of 100 = 1%	$\frac{1}{100} = \frac{1}{100}$
$66\frac{2}{3}$ out of 100 = $66\frac{2}{3}$%	$\frac{66.\overline{6}}{100} = \frac{2}{3}$
75 out of 100 = 75%	$\frac{75}{100} = \frac{3}{4}$

You can write fractions as percents and percents as fractions.

EXAMPLE **Converting a Fraction to a Percent**

Express $\frac{4}{5}$ as a percent.

$\frac{4}{5} = \frac{n}{100}$ • Set up as a proportion.

$5n = 400$ • Solve the proportion.

$n = 80$

$\frac{80}{100} = 80\%$ • Express as a percent.

So, $\frac{4}{5} = 80\%$.

Check It Out

Convert each fraction to a percent.

1 $\frac{3}{5}$ **2** $\frac{3}{10}$ **3** $\frac{9}{10}$ **4** $2\frac{2}{25}$

Changing Percents to Fractions

To change from a percent to a fraction, you write a fraction with the percent as the numerator and 100 as the denominator and then express the fraction in simplest form.

EXAMPLE   **Changing a Percent to a Fraction**

Express 35% as a fraction.

- Change the percent directly to a fraction with a denominator of 100. The number of the percent becomes the numerator of the fraction.

$$35\% = \frac{35}{100}$$

- Simplify, if possible.

$$\frac{35}{100} = \frac{7}{20}$$

35% expressed as a fraction is $\frac{7}{20}$.

 Check It Out

Change each percent to a fraction in simplest form.

5 17%   **6** 5%   **7** 36%   **8** 64%

Changing Mixed Number Percents to Fractions

To change a mixed number percent to a fraction, first change the mixed number to an *improper fraction* (p. 101).

Express $25\frac{1}{2}\%$ as a fraction.

- Change the mixed number to an improper fraction. $25\frac{1}{2}\% = \frac{51}{2}\%$

- Multiply the percent by $\frac{1}{100}$. $\frac{51}{2} \times \frac{1}{100} = \frac{51}{200}$

- Simplify, if possible. $25\frac{1}{2}\% = \frac{51}{200}$

Percents and Decimals

Percents can be expressed as decimals, and decimals can be expressed as percents. *Percent* means part of a hundred or hundredths.

EXAMPLE   Changing Decimals to Percents

Express 0.7 as a percent.

$0.7 \times 100 = 70$ • Multiply the decimal by 100.

$0.7 \longrightarrow 70\%$ • Add the percent sign.

So, 0.7 expressed as a percent is 70%.

A Shortcut for Changing Decimals to Percents

Change 0.7 to a percent.

• Move the decimal point two places to the right. Add zeros, if necessary.

    $0.7 \longrightarrow 70.$

• Add the percent sign.

    $70. \longrightarrow 70\%$

 Check It Out

Write each decimal as a percent.

12 0.27

13 0.007

14 0.018

15 1.5

You can convert percents directly to decimals.

A Shortcut for Changing Percents to Decimals

Change 7% to a decimal.

- Move the decimal point two places to the left.

7% → .7.

- Add zeros, if necessary.

7% = 0.07

 Check It Out

Express each percent as a decimal.

 49%   3%   **18** 180%   0.7%

Fractions and Decimals

A fraction can be written as either a **terminating** or a **repeating decimal**.

Fractions	Decimals	Terminating or Repeating
$\frac{1}{2}$	0.5	terminating
$\frac{1}{3}$	0.3333333 …	repeating
$\frac{1}{6}$	0.166666 …	repeating
$\frac{3}{5}$	0.6	terminating

EXAMPLE **Changing Fractions to Decimals**

Write $\frac{4}{5}$ as a decimal.

$4 \div 5 = 0.8$

- Divide the numerator of the fraction by the denominator.

The remainder is zero. The decimal 0.8 is a *terminating decimal*.

Write $\frac{1}{3}$ as a decimal.

$1 \div 3 = 0.3333333 \ldots$

- Divide the numerator of the fraction by the denominator.

The decimal 0.3333333 is a *repeating decimal*.

$0.\overline{3}$

- Place a bar over the digit that repeats.

So, $\frac{1}{3} = 0.\overline{3}$.

 Check It Out

Use a calculator to find a decimal for each fraction.

20 $\frac{3}{10}$ **21** $\frac{7}{8}$ **22** $\frac{1}{11}$

Changing Decimals to Fractions

Write a decimal as a fraction or mixed number.

- Write the decimal as a fraction with 100 as the denominator.

$0.26 = \frac{26}{100}$

- Express the fraction in simplest form.

$\frac{26}{100} = \frac{13}{50}$

 Check It Out

Write each decimal as a fraction.

23 0.78   **24** 0.54   **25** 0.24

Comparing and Ordering Rational Numbers

A **rational number** is any number that can be written as the quotient of two integers where the divisor is not zero.

rational number	7	$\frac{7}{1}$
rational number	0.4	$\frac{4}{10}$

EXAMPLE Ordering Rational Numbers

Order these rational numbers from least to greatest. You can order rational numbers by writing each as a decimal and then comparing the decimals.

$\frac{2}{3}$, 75%, 0.12

- Change $\frac{2}{3}$ to a decimal.

    $2 \div 3 = 0.67$

- Change 75% to a decimal.

    $75\% = 0.75$

The rational numbers in order from least to greatest are:
0.12, $\frac{2}{3}$, 75%.

Check It Out

Order the rational numbers from least to greatest.

26 $\frac{5}{7}$, 0.23, 62%

27 18%, 0.78, $\frac{5}{9}$

28 0.25, $\frac{1}{2}$, 60%

A corporation raises money by selling stocks—certificates that represent shares of ownership in the corporation. The stock page of a newspaper lists the high, low, and ending prices of stocks for the previous day. It also shows the overall fractional amount by which the prices changed. A (+) sign indicates that the value of a stock increased; a (−) sign indicates that the value decreased.

Suppose that you see a listing on the stock page showing that the closing price of a stock was $21.75 with +0.25 next to it. What do those numbers mean? First, the number tells you that the price of the stock was $21.75. The +0.25 indicates the increase in price from the day before. The stock went up 25¢.

To find the percent increase in the price of the stock, you have to first determine the original price of the stock. The stock went up $0.25, so the original price is $21.75 − $0.25 = $21.50. What is the percent of increase to the nearest whole percent? See HotSolutions for answer.

2·9 Exercises

Change each fraction to a percent.

1. $\dfrac{3}{12}$ 2. $\dfrac{13}{20}$ 3. $\dfrac{63}{100}$ 4. $\dfrac{11}{50}$ 5. $\dfrac{7}{20}$

Change each percent to a fraction in simplest form.

6. 24%   7. 62%   8. 33%

9. 10%   10. 85%

Write each decimal as a percent.

11. 0.6   12. 0.33   13. 0.121

14. 0.64   15. 2.5

Write each percent as a decimal.

16. 27%   17. 14.5%   18. 17%

19. 3%   20. 27.4%

Change each fraction to a decimal. Use a bar to show repeating decimals.

21. $\dfrac{5}{16}$ 22. $\dfrac{2}{3}$ 23. $\dfrac{3}{8}$

24. $\dfrac{2}{5}$ 25. $\dfrac{1}{25}$

Write each decimal as a fraction.

26. 0.76   27. 0.88   28. 0.9

29. 2.5   30. 0.24

Order from least to greatest.

31. $\dfrac{3}{8}$, 0.13, 74%   32. 58%, $\dfrac{4}{5}$, 0.15

33. One survey at Jefferson Middle School reported that 45% of the seventh-grade students wanted the spring dance to be a semiformal. Another survey reported that $\dfrac{9}{20}$ of the seventh-grade students wanted the spring dance to be a semiformal. Could both surveys be correct? Explain.

34. Price Savers is advertising 40% off the $129 in-line skates. Bottom Dollar is advertising the same in-line skates at $\dfrac{1}{4}$ off the price of $129. Which is the better buy? How much would you save with the better buy?

Fractions, Decimals, and Percents

What have you learned?

You can use the problems and the list of words that follow to see what you learned in this chapter. You can find out more about a particular problem or word by referring to the topic number (*for example,* Lesson 2·2).

Problem Set

1. Which fraction is equivalent to $\frac{15}{20}$? (Lesson 2·1)

A. $\frac{3}{4}$ **B.** $\frac{6}{6}$ **C.** $\frac{4}{5}$ **D.** $\frac{5}{4}$

2. Which fraction is greater, $\frac{1}{13}$ or $\frac{3}{21}$? (Lesson 2·2)

3. Write the improper fraction $\frac{17}{9}$ as a mixed number. (Lesson 2·2)

Add or subtract. Write your answers in simplest form. (Lesson 2·3)

4. $\frac{4}{5} + \frac{2}{7}$ **5.** $3\frac{1}{8} - \frac{1}{2}$ **6.** $4 - 2\frac{1}{8}$

Multiply or divide. (Lesson 2·3)

7. $\frac{3}{5} \times \frac{5}{8}$ **8.** $\frac{5}{7} \div 3\frac{1}{2}$ **9.** $2\frac{3}{4} \times \frac{5}{9}$

Find each answer. (Lesson 2·6)

10. $11.44 + 2.834$ **11.** $12.5 - 1.09$

12. 8.07×5.6 **13.** $0.792 \div 0.22$

Solve. Round to the nearest tenth. (Lesson 2·8)

14. What percent of 115 is 40?

15. Find 22% of 66.

16. 8 is 32% of what number?

17. Of the 12 girls on the basketball team, 8 play regularly. What percent play regularly? (Lesson 2·8)

Find the discount and sale price. (Lesson 2·8)

18. regular price: $45, discount percent: 20%

19. regular price: $90, discount percent: 15%

Find the final cost of each purchase. (Lesson 2·8)

20. 6% tax on $94.34

21. 20% tax on $1,500.60

Write each decimal as a percent. (Lesson 2·9)

22. 0.85

23. 0.04

Write each fraction as a percent. (Lesson 2·9)

24. $\frac{1}{100}$

25. $\frac{160}{100}$

Write each percent as a decimal. (Lesson 2·9)

26. 15%

27. 5%

Write each percent as a fraction in simplest form. (Lesson 2·9)

28. 80%

29. 8%

Order from least to greatest. (Lesson 2·9)

30. $\frac{3}{8}$, 0.15, 60%

31. 24%, $\frac{2}{5}$, 0.35

HotWords

Write definitions for the following words.

benchmark (Lesson 2·7)

common denominator
(Lesson 2·1)

compatible numbers
(Lesson 2·6)

cross product (Lesson 2·1)

denominator (Lesson 2·1)

discount (Lesson 2·8)

equivalent (Lesson 2·1)

equivalent fractions (Lesson 2·1)

estimate (Lesson 2·6)

factor (Lesson 2·4)

fraction (Lesson 2·1)

greatest common factor
(Lesson 2·1)

improper fraction (Lesson 2·1)

least common multiple
(Lesson 2·1)

mixed number (Lesson 2·1)

numerator (Lesson 2·1)

percent (Lesson 2·7)

product (Lesson 2·4)

ratio (Lesson 2·7)

rational number (Lesson 2·9)

reciprocal (Lesson 2·4)

repeating decimal (Lesson 2·9)

terminating decimal
(Lesson 2·9)

HotTopic 3

Powers and Roots

What do you know?

You can use the problems and the list of words that follow to see what you already know about this chapter. The answers to the problems are in **HotSolutions** at the back of the book, and the definitions of the words are in **HotWords** at the front of the book. You can find out more about a particular problem or word by referring to the topic number (*for example*, Lesson 3·2).

Problem Set

Write each multiplication, using an exponent. (Lesson 3·1)

1. $3 \times 3 \times 3 \times 3 \times 3$
2. $n \times n \times n$
3. $9 \times 9 \times 9$
4. $x \times x \times x \times x \times x \times x \times x \times x \times x \times x \times x \times x$

Evaluate each square. (Lesson 3·1)

5. 3^2　　　6. 7^2　　　7. 4^2　　　8. 8^2

Evaluate each cube. (Lesson 3·1)

9. 3^3　　10. 4^3　　11. 6^3　　12. 9^3

Evaluate each power of 10. (Lesson 3·1)

13. 10^4　　　　　　14. 10^6
15. 10^{10}　　　　　16. 10^8

Evaluate each square root. (Lesson 3·2)

17. $\sqrt{25}$　　18. $\sqrt{64}$　　19. $\sqrt{100}$　　20. $\sqrt{81}$

Estimate each square root between two consecutive numbers. (Lesson 3·2)

21. $\sqrt{31}$

22. $\sqrt{10}$

23. $\sqrt{73}$

24. $\sqrt{66}$

Use a calculator to find each square root to the nearest thousandth. (Lesson 3·2)

25. $\sqrt{48}$

26. $\sqrt{57}$

27. $\sqrt{89}$

28. $\sqrt{98}$

Write each number in scientific notation. (Lesson 3·3)

29. 36,000,000

30. 600,000

31. 80,900,000,000

32. 540

Write each number in standard form. (Lesson 3·3)

33. 5.7×10^{6}

34. 1.998×10^{3}

35. 7×10^{8}

36. 7.34×10^{5}

HotWords

area (Lesson 3·1)
base (Lesson 3·1)
cube of a number (Lesson 3·1)
exponent (Lesson 3·1)
factor (Lesson 3·1)
perfect square (Lesson 3·2)

power (Lesson 3·1)
scientific notation (Lesson 3·3)
square of a number (Lesson 3·1)
square root (Lesson 3·2)
volume (Lesson 3·1)

3·1 Powers and Exponents

Exponents

You can use multiplication to show repeated addition: $4 \times 2 = 2 + 2 + 2 + 2$. A shortcut that shows the repeated multiplication $2 \times 2 \times 2 \times 2$ is to write the **power** 2^4. The 2, the factor to be multiplied, is called the **base**. The 4 is the **exponent**, which tells you how many times the base is to be multiplied. The expression can be read as "two to the fourth power." When you write an exponent, it is written slightly higher than the base and the size is usually a little smaller.

EXAMPLE **Multiplication Using Exponents**

Write the multiplication $8 \times 8 \times 8 \times 8 \times 8 \times 8$ using an exponent.

• Check that the same factor is being used in the multiplication.

  All the factors are 8.

• Count the number of times 8 is being multiplied.

  There are 6 factors of 8.

• Write the multiplication using an exponent.

  Because the factor 8 is being multiplied 6 times, write 8^6.

So, $8 \times 8 \times 8 \times 8 \times 8 \times 8 = 8^6$.

 Check It Out

Write each multiplication as a power.

❶ $3 \times 3 \times 3 \times 3$

❷ $7 \times 7 \times 7 \times 7$

❸ $a \times a \times a \times a \times a \times a$

❹ $z \times z \times z \times z \times z$

Evaluating the Square of a Number

When a square is made from a segment with length 3, the **area** of the square is $3 \times 3 = 3^2 = 9$. The **square** of 3, then, is 3^2.

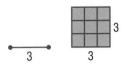

To evaluate a square, identify 3 as the base and 2 as the exponent. Remember that the exponent tells you how many times to use the base as a factor. So 5^2 means to use 5 as a factor 2 times:

$5^2 = 5 \times 5 = 25$

The expression 5^2 can be read as "five to the second power." It can also be read as "5 squared."

EXAMPLE Evaluating the Square of a Number

Evaluate 8^2.

- Identify the base and the exponent.

    The base is 8 and the exponent is 2.

- Write the expression as a multiplication.

    $8^2 = 8 \times 8$

- Evaluate.

    $8 \times 8 = 64$

So, $8^2 = 64$.

 Check It Out

Evaluate each square.

5 4^2

6 8^2

7 3 squared

8 10 squared

Evaluating the Cube of a Number

Evaluating cubes is very similar to evaluating squares. For example, if you want to evaluate 2^3, 2 is the base and 3 is the exponent. Remember that the exponent tells you how many times to use the base as a factor. So, 2^3 means to use 2 as a factor 3 times:

$$2^3 = 2 \times 2 \times 2 = 8$$

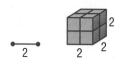

When a cube has edges of length 2, the volume of the cube is $2 \times 2 \times 2 = 2^3 = 8$. The cube of 2, then, is 2^3. The expression 2^3 can be read as "two to the third power." It can also be read as "two cubed."

EXAMPLE Evaluating the Cube of a Number

Evaluate 4^3.

- Identify the base and the exponent.

  The base is 4 and the exponent is 3.

- Write the expression as a multiplication.

  $4^3 = 4 \times 4 \times 4$

- Evaluate.

  $4 \times 4 \times 4 = 64$

So, $4^3 = 64$.

 Check It Out

Evaluate each cube.

9 5^3

10 10^3

11 8 cubed

12 6 cubed

Powers of Ten

Our decimal system is based on 10. For each factor of 10, the decimal point moves one place to the right.

$$2.\underset{\smile}{11} \longrightarrow 21.1 \qquad 19.\underset{\smile}{05} \longrightarrow 1,905 \qquad 7.\underset{\smile}{} \longrightarrow 70$$
$$\times\ 10 \qquad\qquad\quad \times\ 100 \qquad\qquad \times\ 10$$

When the decimal point is at the end of a number and the number is multiplied by 10, a zero is added at the end of the number.

Try to discover a pattern for the powers of 10.

Power	As a Multiplication	Result	Number of Zeros
10^3	$10 \times 10 \times 10$	1,000	3
10^5	$10 \times 10 \times 10 \times 10 \times 10$	100,000	5
10^7	$10 \times 10 \times 10 \times 10 \times 10 \times 10 \times 10$	10,000,000	7
10^9	$10 \times 10 \times 10 \times 10 \times 10 \times 10 \times 10 \times 10 \times 10$	1,000,000,000	9

Notice that the number of zeros after the 1 is the same as the power of 10. This means that if you want to evaluate 10^7, you simply write a 1 followed by 7 zeros: 10,000,000.

Check It Out

Evaluate each power of 10.

13 10^2

14 10^6

15 10^8

16 10^3

APPLICATION When Zeros Count

Usually you think that a zero means "nothing." But when zeros are related to a power of 10, you can get some fairly large numbers. A billion is the name for 1 followed by 9 zeros; a quintillion is the

name for 1 followed by 18 zeros. You can write out all the zeros or use mathematical shorthand for these numbers.

$$1 \text{ billion} = 1,000,000,000 \text{ or } 10^9$$

$$1 \text{ quintillion} = 1,000,000,000,000,000,000 \text{ or } 10^{18}$$

What name would you use for 1 followed by one hundred zeros? According to the story, when the mathematician Edward Kasner asked his nine-year-old nephew to think up a name for this number, his nephew called it a *googol*. And that's the name used for 10^{100} today.

Suppose that you could count at the rate of 1 number each second. If you started counting now and continued to count for 12 hours per day, it would take about 24 days to count to 1 million (1,000,000 or 10^6). Do you think you could count to a googol in your lifetime? See **HotSolutions** for the answers.

3·1 Exercises

Write each multiplication using an exponent.

1. $8 \times 8 \times 8$

2. $3 \times 3 \times 3 \times 3 \times 3 \times 3 \times 3$

3. $y \times y \times y \times y \times y \times y$

4. $n \times n \times n \times n \times n \times n \times n \times n \times n$

5. 15×15

Evaluate each square.

6. 5^2

7. 14^2

8. 7^2

9. 1 squared

10. 20 squared

Evaluate each cube.

11. 5^3

12. 9^3

13. 11^3

14. 3 cubed

15. 8 cubed

Evaluate each power of 10.

16. 10^2

17. 10^{14}

18. 10^6

19. What is the area of a square with side lengths of 9?

    A. 18   **B.** 36

    C. 81   **D.** 729

20. What is the volume of a cube with side lengths of 5?

    A. 60   **B.** 120

    C. 125   **D.** 150

3·2 Square Roots

Square Roots

In mathematics, certain operations are opposites; that is, one operation "undoes" the other. Addition undoes subtraction: $11 - 3 = 8$, so $8 + 3 = 11$. Multiplication undoes division: $16 \div 8 = 2$, so $2 \times 8 = 16$.

The opposite of squaring a number is finding the **square root**. You know that $3^2 = 9$. The square root of 9 is the number that is multiplied by itself to get 9, which is 3. The symbol for square root is $\sqrt{\ }$. Therefore, $\sqrt{9} = 3$.

EXAMPLE	Finding the Square Root

Find $\sqrt{25}$.

• Think, what number times itself makes 25?

$5 \times 5 = 25$

• Find the square root.

Because $5 \times 5 = 25$, the square root of 25 is 5.

So, $\sqrt{25} = 5$.

 Check It Out

Find each square root.

1 $\sqrt{9}$

2 $\sqrt{36}$

3 $\sqrt{81}$

4 $\sqrt{121}$

Estimating Square Roots

The table shows the first ten **perfect squares** and their square roots.

Perfect square	1	4	9	16	25	36	49	64	81	100
Square root	1	2	3	4	5	6	7	8	9	10

Estimate $\sqrt{50}$. In this problem, 50 is the square. In the table, 50 lies between 49 and 64. So, $\sqrt{50}$ must be between $\sqrt{49}$ and $\sqrt{64}$, which is between 7 and 8. To estimate the value of a square root, you can find the two consecutive numbers that the square root must be between.

EXAMPLE Estimating a Square Root

Estimate $\sqrt{95}$.

- Identify the perfect squares that 95 is between.

    95 is between 81 and 100.

- Find the square roots of the perfect squares.

    $\sqrt{81} = 9$ and $\sqrt{100} = 10$.

- Estimate the square root.

    $\sqrt{95}$ is between 9 and 10.

 Check It Out

Estimate each square root.

5 $\sqrt{42}$

6 $\sqrt{21}$

7 $\sqrt{5}$

8 $\sqrt{90}$

Better Estimates of Square Roots

To find a more accurate estimate for the value of a square root, you can use a calculator. Most calculators have a $\boxed{\sqrt{}}$ key for finding square roots.

On some calculators, the $\sqrt{}$ function is shown not on a key but above the $\boxed{x^2}$ key on the calculator's key pad. If this is true for your calculator, you should also find a key that has either $\boxed{\text{INV}}$ or $\boxed{\text{2nd}}$ on it.

To use the $\sqrt{}$ function, you press $\boxed{\text{INV}}$ or $\boxed{\text{2nd}}$ and then the key with $\sqrt{}$ above it.

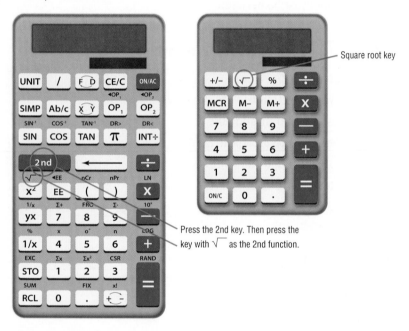

Square root key

Press the 2nd key. Then press the key with $\sqrt{}$ as the 2nd function.

The estimate of the square root of a number that is not a perfect square will be a decimal, and the entire calculator display will be used. Generally, you should round square roots to the nearest thousandth. (Remember that the thousandths place is the third place after the decimal point.)

See Lessons 8•1 and 8•2 for more about calculators.

Use a calculator to find $\sqrt{34}$ to the nearest thousandth.

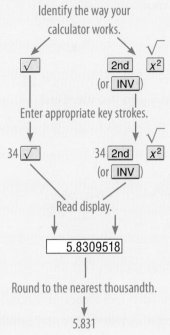

So, $\sqrt{34} = 5.831$.

 Check It Out

Use a calculator to find each square root to the nearest thousandth.

9 $\sqrt{3}$

10 $\sqrt{47}$

11 $\sqrt{86}$

12 $\sqrt{97}$

3·2 Exercises

Find each square root.

1. $\sqrt{16}$ 2. $\sqrt{49}$

3. $\sqrt{144}$ 4. $\sqrt{25}$

5. $\sqrt{100}$

6. $\sqrt{31}$ is between which two numbers?
   A. 3 and 4
   B. 5 and 6
   C. 29 and 31
   D. None of these

7. $\sqrt{86}$ is between which two numbers?
   A. 4 and 5
   B. 8 and 9
   C. 9 and 10
   D. 83 and 85

8. $\sqrt{23}$ is between what two consecutive numbers?

9. $\sqrt{50}$ is between what two consecutive numbers?

10. $\sqrt{112}$ is between what two consecutive numbers?

Use a calculator to find each square root to the nearest thousandth.

11. $\sqrt{2}$ 12. $\sqrt{20}$

13. $\sqrt{50}$ 14. $\sqrt{75}$

15. $\sqrt{1,000}$ 16. $\sqrt{8,100}$

3·3 Scientific Notation

Using Scientific Notation

When you work with very large numbers, such as 829,000,000, it can be difficult to keep track of the zeros. You can express such numbers in **scientific notation** by writing the number as a product of a factor and a power of ten (p. 163).

Scientific notation is also used to express very small numbers. Study the pattern of products below.

$$1.25 \times 10^2 = 125$$
$$1.25 \times 10^1 = 12.5$$
$$1.25 \times 10^0 = 1.25$$
$$1.25 \times 10^{-1} = 0.125$$
$$1.25 \times 10^{-2} = 0.0125$$
$$1.25 \times 10^{-3} = 0.00125$$

Notice that multiplying by a negative power of 10 moves the decimal point to the left the same number of places as the absolute value of the exponent.

EXAMPLE **Writing a Very Large Number in Scientific Notation**

Write 4,250,000,000 in scientific notation.

4.250000000.

- Move the decimal point so that only one digit is to the left of the decimal.

4.250000000.
9 places

- Count the number of decimal places that the decimal has to be moved to the left.

4.25×10^9

- Write the number without the ending zeros, and multiply by the correct power of 10.

So, 4,250,000,000 can be written 4.25×10^9.

Check It Out

Write each number in scientific notation.

1. 53,000
2. 4,000,000
3. 70,800,000,000
4. 26,340,000

Converting from Scientific Notation to Standard Form

Remember that each factor of 10 moves the decimal point one place to the right. When the last digit of the number is reached, there may still be some factors of 10 remaining. Add a zero at the end of the number for each remaining factor of 10.

EXAMPLE Converting to a Number in Standard Form

Write 7.035×10^6 in standard form.

• The exponent tells how many places to move the decimal point.

    The decimal point moves to the right 6 places.

• Move the decimal point the exponent's number of places to the right. Add zeros at the end of the number if needed.

$$7.035000.$$

Move the decimal point
to the right 6 places.

• Write the number in standard form.

So, 7.035×10^6 is 7,035,000 in standard form.

Check It Out

Write each number in standard form.

5. 6.7×10^4
6. 2.89×10^8
7. 1.703×10^5
8. 8.52064×10^{12}

3·3 Exercises

Write each number in scientific notation.

1. 630,000
2. 408,000,000
3. 80,000,000
4. 15,020,000,000,000
5. 350
6. 7,060
7. 10,504,000
8. 29,000,100,000,000,000

Write each number in standard form.

9. 4.2×10^7
10. 5.71×10^4
11. 8.003×10^{10}
12. 5×10^8
13. 9.4×10^2
14. 7.050×10^3
15. 5.0203×10^9
16. 1.405×10^{14}

17. Which of the following expresses the number 5,030,000 in scientific notation?
    A. 5×10^6 B. 5.03×10^6
    C. 50.3×10^5 D. None of these
18. Which of the following expresses the number 3.09×10^7 in standard form?
    A. 30,000,000   B. 30,900,000
    C. 3,090,000,000   D. None of these
19. When written in scientific notation, which of the following numbers will have the greatest power of 10?
    A. 93,000   B. 408,000   C. 5,556,000   D. 100,000,000
20. In scientific notation, what place value does 10^6 represent?
    A. hundred thousands   B. millions
    C. ten millions   D. hundred millions

Powers and Roots

What have you learned?

You can use the problems and the list of words that follow to see what you learned in this chapter. You can find out more about a particular problem or word by referring to the topic number (*for example,* Lesson 3·2).

Problem Set

Write each multiplication, using an exponent. (Lesson 3·1)

1. $20 \times 20 \times 20 \times 20 \times 20$

2. $k \times k \times k \times k \times k \times k \times k$

3. $4 \times 4 \times 4 \times 4 \times 4 \times 4$

4. $y \times y$

Evaluate each square. (Lesson 3·1)

5. 4^2 **6.** 8^2 **7.** 13^2 **8.** 4^2

Evaluate each cube. (Lesson 3·1)

9. 11^3 **10.** 12^3 **11.** 22^3 **12.** 8^3

Evaluate each power of 10. (Lesson 3·1)

13. 10^3 **14.** 10^6

15. 10^8 **16.** 10^{12}

Evaluate each square root. (Lesson 3·2)

17. $\sqrt{4}$ **18.** $\sqrt{36}$ **19.** $\sqrt{144}$ **20.** $\sqrt{100}$

Estimate each square root between two consecutive numbers. (Lesson 3·2)

21. $\sqrt{55}$ **22.** $\sqrt{19}$

23. $\sqrt{99}$ **24.** $\sqrt{14}$

Use a calculator to find each square root to the nearest thousandth. (Lesson 3·2)

25. $\sqrt{50}$

26. $\sqrt{18}$

27. $\sqrt{73}$

28. $\sqrt{7}$

Write each number in scientific notation. (Lesson 3·3)

29. 2,902,000

30. 113,020

31. 40,100,000,000,000

32. 8,060

Write each number in standard form. (Lesson 3·3)

33. 1.02×10^3

34. 2.701×10^7

35. 3.01×10^{12}

36. 6.1×10^2

HotWords

Write definitions for the following words.

area (Lesson 3·1)

base (Lesson 3·1)

cube of a number (Lesson 3·1)

exponent (Lesson 3·1)

factor (Lesson 3·1)

perfect square (Lesson 3·2)

power (Lesson 3·1)

scientific notation (Lesson 3·3)

square of a number (Lesson 3·1)

square root (Lesson 3·2)

volume (Lesson 3·1)

HotTopic 4

Data, Statistics, and Probability

What do you know?

You can use the problems and the list of words that follow to see what you already know about this chapter. The answers to the problems are in **HotSolutions** at the back of the book, and the definitions of the words are in **HotWords** at the front of the book. You can find out more about a particular problem or word by referring to the topic number (*for example,* Lesson 4·2).

Problem Set

1. Sonja walked through her neighborhood and stopped at her friends' houses to ask whether they planned to vote. Is this a random sample? (Lesson 4·1)

2. Rebecca asked people in the mall the following question: What do you think of the ugly new city hall building? Was her question biased or unbiased? (Lesson 4·2)

For Exercises 3 and 4, use the following data about the number of minutes spent doing sit-ups. (Lesson 4·2)

3 4 2 4 1 3 1 1 4 3 4 5 2 6 9 1

3. Make a line plot of the data.

4. Make a histogram of the data.

5. Hani had the following scores on his math tests: 90, 85, 88, 78, and 96. What is the range of scores? (Lesson 4·4)

6. Find the mean and median of the scores in Exercise 5.

7. $P(5, 1) = $ ____ (Lesson 4·5)

8. $C(8, 2) = $ ____ (Lesson 4·5)

9. What is the value of 7!? (Lesson 4·5)

Use the following information to answer Exercises 10 and 11.
(Lesson 4·5)

A bag contains 20 marbles—8 red, 6 blue, and 6 yellow.

10. One marble is drawn at random. What is the probability that it is blue or yellow?

11. Two marbles are drawn at random. What is the probability that both are white?

HotWords

average (Lesson 4·4)	**percent** (Lesson 4·2)
circle graph (Lesson 4·2)	**permutation** (Lesson 4·5)
combination (Lesson 4·5)	**population** (Lesson 4·1)
correlation (Lesson 4·3)	**probability** (Lesson 4·5)
dependent events (Lesson 4·5)	**probability line** (Lesson 4·5)
double-bar graph (Lesson 4·2)	**random sample** (Lesson 4·1)
event (Lesson 4·5)	**range** (Lesson 4·4)
experimental probability (Lesson 4·5)	**sample** (Lesson 4·1)
	scatter plot (Lesson 4·3)
factorial (Lesson 4·5)	**simple event** (Lesson 4·5)
histogram (Lesson 4·2)	**spinner** (Lesson 4·5)
independent event (Lesson 4·5)	**stem** (Lesson 4·2)
leaf (Lesson 4·2)	**stem-and-leaf plot** (Lesson 4·2)
line graph (Lesson 4·2)	**survey** (Lesson 4·1)
mean (Lesson 4·4)	**table** (Lesson 4·1)
median (Lesson 4·4)	**tally mark** (Lesson 4·1)
mode (Lesson 4·4)	**theoretical probability** (Lesson 4·5)
outcome (Lesson 4·5)	
outcome grid (Lesson 4·5)	**tree diagram** (Lesson 4·5)

4·1 Collecting Data

Surveys

Have you ever been asked to name your favorite movie or what kind of pizza you like? These kinds of questions are often asked in **surveys**. A statistician studies a group of people or objects, called a **population**. Statisticians usually collect information from a small part of the population, called a **sample**.

In a survey, 200 seventh graders in Lakeville were chosen at random and asked what their favorite subject is in school. The following bar graph shows the percent of students who identified each favorite subject.

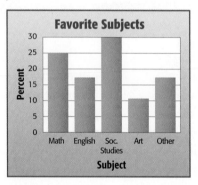

In this case, the population is all seventh graders in Lakeville. The sample is the 200 students who were actually asked to identify their favorite subjects.

In any survey:

• The population consists of the people or objects about which information is desired.

• The sample consists of the people or objects in the population that are actually studied.

Identify the population and the size of the sample:

1 500 adults who were registered to vote were surveyed at random to find out whether they were in favor of increasing taxes to pay for education.

2 200 fish in Sunshine Lake were captured, tagged, and released.

Random Samples

When you choose a sample to survey, make sure that the sample is representative of the population. You should also make sure that it is a **random sample**, where each person in the population has an equal chance of being included.

Ms. Landover wants to find out whether her students want to go to a science museum, a fire station, the zoo, or a factory on a field trip. To identify a sample of the population, she writes the names of her students on cards and randomly draws 20 cards from a bag. She then asks those 20 students where they want to go on the field trip.

EXAMPLE Finding a Random Sample

Determine whether Ms. Landover's sample is random.

- Define the population.

    The population is Ms. Landover's class.

- Define the sample.

    The sample consists of 20 students.

- Determine whether the sample is random.

Because every student in Ms. Landover's class has the same chance of being chosen, the sample is random.

Check It Out

3. Suppose that you ask 20 people in a grocery store which grocery store is their favorite. Is the sample random?

4. A student assigns numbers to his 24 classmates and then writes the numbers on slips of paper. He draws 10 numbers and asks those students to identify their favorite TV program. Is the sample random?

5. How do you think you could select a random sample of your classmates?

Biased Samples

A sample must be chosen very carefully to achieve valid results. An **unbiased sample** is selected so that it is representative of the entire population. In a **biased sample**, one or more parts of the population receive greater representation than others.

A *convenience sample* is a biased sample. This sampling is made up of members of a population that are easily accessible. For example: Your assignment is to conduct a survey to find the favorite kinds of food among seventh graders. But you do not have the survey ready for the next class, so you survey the members of the class you are in—gym class. This sample is biased because only members of your gym class are represented.

Another type of biased sample is a *voluntary-response sample*. A voluntary-response sample includes only those who want to participate in the sampling. For example: You ask volunteers to participate in a survey about global warming. The students who respond to the request might want to express their opinions about the topic. The data you collect will be biased.

COLLECTING DATA

4•1

Determine whether the sample is biased or unbiased. Determine the validity of the results.

- To determine the kind of snacks people like to eat, every tenth person who walks past a park entrance is surveyed. Is the data biased or unbiased?

Because the survey is random, the sample is unbiased and the conclusion is valid.

An online newspaper asks its visitors to indicate their preference for a presidential candidate. Is the data biased or unbiased?

The results of a voluntary response sample do not necessarily represent the entire population. The sample is biased, and the conclusion is not valid.

Check It Out

Tell whether the sample is biased or unbiased.

6 A survey of 100 randomly selected households in Florida was conducted to determine the number of winter coats the average person has.

7 To determine how many people bring their lunch to work, a researcher randomly samples employees from corporations across the country.

Questionnaires

When you write questions for a survey, it is important to make sure that the questions are not biased; that is, the questions should not assume anything or influence the answers.

The two questionnaires on page 182 are designed to find out what kinds of books your classmates like and what they like to do in the summer. The first questionnaire uses biased questions. The second questionnaire uses questions that are not biased.

Questionnaire 1:

    A. What mystery novel is your favorite?

    B. What TV programs do you like to watch in the summer?

Questionnaire 2:

    A. What kind of books do you like to read?

    B. What do you like to do in the summer?

Develop a questionnaire:

• Decide what topic you want to ask about.

• Define a population and decide how to select a sample from that population.

• Develop questions that are unbiased.

Check It Out

  8 Why is **A** in Questionnaire 1 biased?

  9 Why is **B** in Questionnaire 2 better than **B** in Questionnaire 1?

  10 Write a question that asks the same thing as the following question but is not biased: Are you an interesting person who reads many books?

Compiling Data

After Ms. Landover collects the data from her students about field trip preferences, she decides how to display the results. As she asks students their preferences, she uses **tally marks** to tally the answers in a table. The following **table** shows their answers.

Preferred Field Trip	Number of Students
Science museum	⦀⦀ |||
Zoo	||||
Fire station	⦀⦀
Factory	|||

To compile data in a table:

• List the categories or questions in the first column or row.

• Tally the responses in the second column or row.

 Check It Out

⓫ How many students chose the zoo?

⓬ Which field trip was chosen by the fewest students?

⓭ If Ms. Landover uses the survey to choose a field trip, which one should she choose? Explain.

Marketing research often involves the use of surveys and product sampling to help predict how well certain products will do. The results of this process might then be used to determine what part of the population is likely to buy these new products.

A company planning to introduce three new flavors of ice cream—Banana Bonanza, Raspberry Rush, and Kiwi Kiss—handed out samples of each flavor to 300 seventh graders. They were asked which flavor or flavors they preferred. The results are displayed in the diagram.

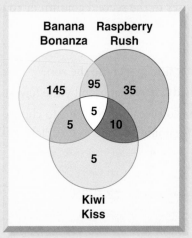

Use the information provided above to predict how many out of thirty seventh graders at your school should prefer only Banana Bonanza. See **HotSolutions** for the answer. Then actually survey thirty seventh graders about the flavor they think they would prefer.

4·1 Exercises

1. Two hundred seventh graders in Carroll Middle School were asked to identify their favorite after-school activity. Identify the population and the sample. How big is the sample?

2. Salvador knocked on 25 doors in his neighborhood. He asked the residents who answered if they were in favor of the idea of closing the ice-skating rink in their city. Is the sample random?

Are the following questions biased? Explain.

3. Are you happy about the lovely plants being planted in the school yard?

4. How many times each week do you eat a school lunch?

Rewrite the questions in an unbiased way.

5. Are you thoughtful about not riding your bike on the sidewalk?

6. Do you like the long walk to school, or do you prefer to take the bus or get a ride?

Mr. Kemmeries asked his students which type of food they wanted at a class party and tallied the following information.

Type of Food	Number of Seventh Graders	Number of Eighth Graders
Chicken fingers	ʜʜʜ IIII	ʜʜʜ ʜʜʜ II
Bagels	ʜʜʜ I	III
Pizza	ʜʜʜ ʜʜʜ II	ʜʜʜ ʜʜʜ
Sandwiches	ʜʜʜ II	ʜʜʜ I
Turkey franks	III	ʜʜʜ I

7. Which type of food was most popular? How many students preferred that type?

8. Which type of food was preferred by 13 students?

9. How many students were surveyed?

4.2 Displaying Data

Interpret and Create a Table

Statisticians collect data about people or objects. One way to show the data is to use a table. Here are the number of letters in each word in the first sentence of *Little House in the Big Woods*.

4 4 1 4 5 5 3 1 6 4 5 2 3 3 5 2 9 2 1 6 4 5 4 2 4

EXAMPLE   Making a Table

Make a table to organize the data about the number of letters in the words.

• Name the first column or row *what* you are counting.

   Label the first row *Number of Letters*.

• Tally the amounts for each category in the second column or row.

Number of Letters	1	2	3	4	5	6	7	8	9
Number of Words	III	IIII	III	₩₩ II	₩₩	II			I
Frequency	3	4	3	7	5	2	0	0	1

• Count the tallies, and record the number in the second row.

Number of Letters	1	2	3	4	5	6	7	8	9
Number of Words	3	4	3	7	5	2	0	0	1

The most common number of letters in a word was 4. Three words have 1 letter.

 Check It Out

Use the table above to answer Exercise 1.

❶ Which letter counts are not represented in any words?

❷ Make a table with the following data to show the number of hours seventh graders spent watching TV daily.

9 9 3 7 6 3 3 2 4 0 1 1 0 2 1 2 1 1 4 0 1 0

Interpret and Create a Circle Graph

Another way to show data is to use a **circle graph**. Aisha conducted a survey of her classmates. She found out that 24% walk to school, 32% take the school bus, 20% get a ride, and 24% ride their bikes. She wants to make a circle graph to show her data.

EXAMPLE | Making a Circle Graph

Make a circle graph.

- Determine the **percent** of the whole of data represented by each unit.

    In this case, the percents are given.

- Make a chart. Multiply each percent by 360°, the number of degrees in a circle.

How We Get to School		
Type	**%**	**Central $\angle$ measure**
Bike	24%	$360° \times 24\% = 86.4°$
Bus	32%	$360° \times 32\% = 115.2°$
Get a ride	20%	$360° \times 20\% = 72°$
Walk	24%	$360° \times 24\% = 86.4°$

- Draw a circle, measure each central angle, and complete the graph.

How We Get to School

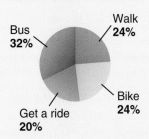

From the graph, you can see that more than half of the students travel by bus or get a ride.

Check It Out

Use the circle graph to answer Exercises 3 and 4.

Animals in Pet Show

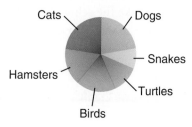

3 About what fraction of the pets were dogs?

4 About what fraction of the pets were not cats or dogs?

5 The following entries were received at the county fair. Make a circle graph to show the entries.

Pies: 76   Jams: 36   Cakes: 50   Bread: 38

Interpret and Create a Line Plot

You have used tally marks to show data. Suppose that you collect the following information about the number of people in your classmates' families.

4 2 3 6 5 6 3 2 4 3 5 5 3 7 5 3 4 3

To make a line plot, place Xs above a number line.

• Draw a number line showing the numbers in your data set.
  Draw a number line showing the numbers 2 through 7.

• Place an X to represent each result above the appropriate number on the number line.
  For this line plot, put an X above the number of people in each family.

• Title the graph. You can call it "Number in Our Families."

DISPLAYING DATA

4•2

Your line plot should look like this:

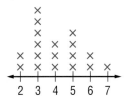

Number in our Families

You can tell from your line plot that the most frequent number of members in a family is 3.

Check It Out

6 What is the greatest number of people in a classmate's family?

7 How many classmates have 5 people in their family?

8 Make a line plot to show the number of letters in each word in the first sentence of *Little House in the Big Woods* (p. 186).

Interpret a Line Graph

You know that a *line graph* can be used to show changes in data over time. The following line graph shows cell phone usage in City X over a seven-year period.

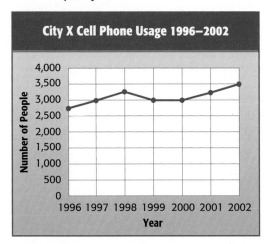

From the graph, you can see that cell phone usage increased, then decreased to level out for two years, and then steadily increased again.

 Check It Out

Use the graph on page 189 to answer Exercises 9–11.

9 What was the year of greatest usage?

10 In what year was a decrease first experienced?

11 *True* or *false*: Between the years 1996 and 2002, there were never more than 3,000 people in the city who used cell phones.

Interpret a Stem-and-Leaf Plot

The following numbers show the number of team wins in one season.

38 23 36 22 30 24 31 26 27 32 27 35 23 40 23 32 25 31
28 28 26 31 28 41 25 29

You know that you could make a table or a line graph to show this information. Another way to show the information is to make a **stem-and-leaf plot**. The following stem-and-leaf plot shows the number of wins.

Team Wins

Stem	Leaf
2	2 3 3 3 4 5 5 6 6 7 7 8 8 8 9
3	0 1 1 1 2 2 5 6 8
4	0 1

2 | 2 = 22 *wins*

Notice that the tens digits appear in the left-hand columns. These are called **stems**. Each digit on the right is called a **leaf**. From looking at the plot, you can tell that most of the teams have from 20 to 29 wins.

Check It Out

Use the stem-and-leaf plot showing the number of books read by students in the summer reading contest at the library to answer the following questions.

12 How many students participated in the contest?

13 How many books did the students read in all?

14 Two students read the same number of books. How many books was that?

Books Read

Stem	Leaf
1	0 2 3
2	0 2 2 4 5 8
3	
4	1 3 4

4 | 1 = 41 books

Interpret and Create a Bar Graph

Another type of graph you can use to show data is called a *bar graph*. In this graph, either horizontal or vertical bars are used to show data. Consider the data showing the tallest building in the United States.

Heights of U.S. Skyscrapers	
Sears Tower	442 m
Empire State Building	381 m
Bank of America Tower	366 m
Aon Center	346 m
John Hancock Center	344 m

You can make a bar graph to show these heights:

• Choose a vertical scale, and decide what to place along the horizontal scale.

In this case, the vertical scale can show meters in increments of 50 meters and the horizontal scale can show the buildings.

• Above each building, draw a bar of the appropriate height.

• Title the graph.

You can call it "Heights of U.S. Skyscrapers."

Your bar graph should look like this.

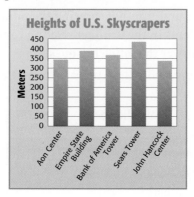

Heights of U.S. Skyscrapers

From the graph, you can see that the Sears Tower is the tallest building.

Check It Out

15 What is the shortest building shown?

Interpret a Double-Bar Graph

You know that you can show information in a bar graph. If you want to show information about two or more things, you can use a **double-bar graph**. This graph shows how many students were on the honor roll at the end of last year and this year.

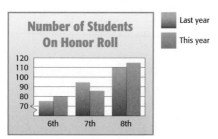

You can see from the graph that more eighth graders were on the honor roll this year than last year.

Check It Out

16 About how many students were on the honor roll at the end of this year?

Interpret and Create a Histogram

A special kind of bar graph that shows frequency of data is called a **histogram**. Suppose that you ask several classmates how many books they have checked out from the school library and collect the following data:

3 3 3 1 1 0 4 2 1 3 4 2 1 0 1 6

EXAMPLE Making a Histogram

Create a histogram.

• Make a frequency table.

Books	Tally	Frequency
0	II	2
1	ͰͰͰ	5
2	II	2
3	IIII	4
4	II	2
5		0
6	I	1

• Make a histogram showing the frequencies.

• Title the graph.

You can call it "Books Checked Out from the Library."

Your histogram will look like this:

You can see from the histogram that no student has five books checked out.

Check It Out

ⓘ How many students did you survey about library books (p. 193)?

⑱ Make a histogram from the data about *Little House in the Big Woods* (p. 186).

APPLICATION And the Winner Is . . .

At the end of February 2007, a performer won three Grammys.

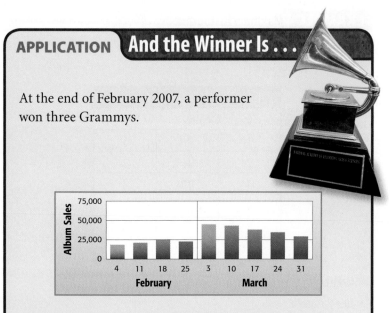

Based on the graph, how did winning the Grammy awards affect album sales? How often were the sales tabulated? What kind of graph is this? See **HotSolutions** for the answers.

4·2 Exercises

1. Make a table and a histogram to show the following data.
   Number of Times Students Purchased Hot Lunch
   4 3 5 1 4 0 3 0 2 3 1 5 4 3 1 5 0 2 3 1

2. How many students were surveyed?

3. Make a line plot to show the data in Exercise 1.

4. Use your line plot to describe the number of hot lunches purchased.

5. Ryan spends $10 of his allowance each week on school lunch and $7 on CDs and magazines. He saves $3. Make a circle graph to show how Ryan spends his allowance, and write a sentence about the graph.

6. The following stem-and-leaf plot shows the number of deer counted each day at a wildlife feeding station over a three-week period.

   Deer at Feeding Station

Stem	Leaf
1	0 2 3 4 4 4 6 8
2	0 0 3 4 4 4 5 6 8 8
3	1 1 2

   1 | 3 = 13

   Draw a conclusion from the plot.

7. Six seventh graders ran laps around the school. Vanessa ran 8 laps, Tomas ran 3, Vedica ran 4, Samuel ran 6, Forest ran 7, and Tanya ran 2. Make a bar graph to show this information.

4·3 Analyzing Data

Scatter Plots

You can analyze and interpret data. A **scatter plot** displays two sets of data on the same coordinate grid. The plot should show whether the data are related. It can also show trends, which may be helpful in making predictions. If the points on a scatter plot come close to lying on a straight line, the two sets of data are related.

Juan collected the following information showing the number of boxes of cookies sold by each person in his town's marching band and the number of years each person had played in band.

Years in Band	2	3	4	1	3	5	2	3	1	3	2	4	1	1
Boxes Sold	35	34	20	36	28	15	20	35	41	17	29	27	43	32

Use the information to make a scatter plot. First, write the data as ordered pairs, and then graph the ordered pairs. Label the vertical and horizontal axes, and title the graph.

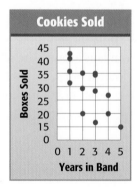

This scatter plot shows an overall downward trend; that is, the longer a person is in the band, the fewer boxes of cookies he or she sells.

Make a scatter plot showing the following data.

1 Relationship between the number of hours slept the night before a test and test scores

Student	1	2	3	4	5	6	7	8	9	10
Hours Slept	7	8	7	6	8	5	7	8	9	7
Test Score	83	87	87	82	63	73	80	81	89	84

2 Mintage of nickels made by year

Year	Number (rounded to nearest million)
1991	600
1992	500
1993	400
1994	700
1995	100
1996	100
1997	500
1998	600
1999	500

Correlation

Suppose that you collect some data about 10 students in your class. You can write the information you have collected as ordered pairs and then plot the ordered pairs. The following scatter plots show the data you collected.

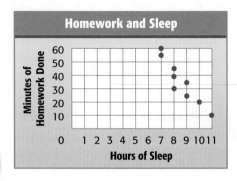

The graph (left) shows a negative **correlation**. The longer students spent on homework, the less sleep they got. A downward trend in points shows a negative correlation.

The graph (right) shows a positive correlation. The longer students spent on homework, the higher their test scores were. An upward trend in points shows a positive correlation.

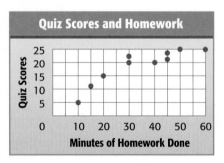

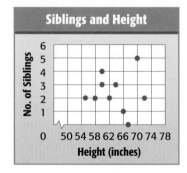

The graph (left) shows no correlation. The number of siblings had no relationship to student height. If there is no trend in graph points, there is no correlation.

4·3

ANALYZING DATA

Check It Out

3 Which of the following scatter plots shows a negative correlation?

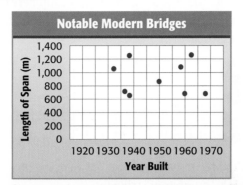

Notable Modern Bridges

Length of Span (m) vs. Year Built

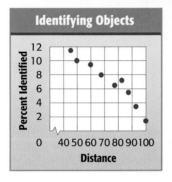

Identifying Objects

Percent Identified vs. Distance

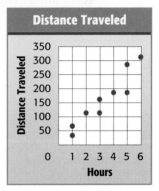

Distance Traveled

Distance Traveled vs. Hours

4 Describe the correlation in the scatter plot showing the relationship between the year a bridge was built and its length.

5 Which scatter plot shows a positive correlation?

4·3 Exercises

1. Make a scatter plot showing the relationship between the number of times at bat and strikeouts for the local high school team.

Times at bat	100	110	102	109	106	101	107	104	100
Strikeouts	20	29	36	28	33	30	27	36	35
Player	1	2	3	4	5	6	7	8	9

2. Describe the relationship, if any, between the number of times at bat and the number of strikeouts.

3. Choose the best answer.

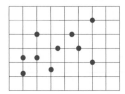

   A. positive correlation
   B. negative correlation
   C. no correlation

Describe the correlation in each of the following scatter plots.

4. 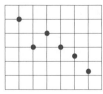   5. 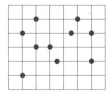   6.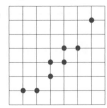

4·4 Statistics

Cleve collected the following data about the amounts of money his friends raised selling cookies.

$10, $10, $12, $15, $18, $19, $21, $23, $25, $28, $32, $35, $38, $39, $140

Cleve said that his classmates typically raise $10, but Kimiko disagreed. She said that the typical amount is $23. A third classmate, Gabe, said that they were both wrong—the typical amount is $31. Each was correct because each was using a different common measure to describe the central tendency of a set of data.

Mean

One measure of the central tendency of a set of data is the **mean**. To find the mean, or **average**, add the amounts the students raised, and divide by the number of amounts.

EXAMPLE **Finding the Mean**

Find the mean amount raised by each friend selling cookies.

- Add the amounts.

    $10 + $10 + $12 + $15 + $18 + $19 + $21 + $23 + $25 + $28 + $32 + $35 + $38 + $39 + $140 = $465

- Divide the total by the number of amounts.

    In this case, there are 15 amounts:

    $465 ÷ 15 = $31

The mean amount raised is $31. Gabe used the mean when he said the typical amount raised is $31.

Find the mean.

1. 14, 12.5, 11.5, 4, 8, 14, 14, 10
2. 45, 100, 82, 83, 100, 46, 90
3. 321, 354, 341, 337, 405, 399
4. Winnie priced jeans at five stores. They were priced at $40, $80, $51, $64, and $25. Find the mean price.

Median

Another measure of central tendency is the *median*. The **median** is the middle number in the data when the numbers are arranged in order. Look again at the amounts raised by the friends selling cookies.

$10, $10, $12, $15, $18, $19, $21, $23, $25, $28, $32, $35, $38, $39, $140

EXAMPLE Finding the Median

Find the median amount of money raised by the friends selling cookies.

- Arrange the data in numerical order from least to greatest or greatest to least.

  Looking at the amounts raised, you can see that they are already arranged in order.

- Find the middle number.

  There are 15 numbers. The middle number is $23 because there are 7 numbers greater than $23 and 7 numbers less than $23.

Kimiko was using the median when she said that the typical amount raised is $23.

When the number of amounts is even, you find the median by finding the mean of the two middle numbers.

EXAMPLE **Finding the Median of an Even Number of Data**

Find the median of 2, 6, 4, 3, 5, and 9.

• Arrange the numbers in order from least to greatest or greatest to least.

    2, 3, 4, 5, 6, 9   or   9, 6, 5, 4, 3, 2

• Find the mean of the two middle numbers.

    The two middle numbers are 4 and 5.

    $(4 + 5) \div 2 = 4.5$

The median is 4.5. Half the numbers are greater than 4.5, and half the numbers are less than 4.5.

Check It Out

Find the median.

5 11, 18, 11, 5, 17, 18, 8

6 5, 9, 2, 6

7 45, 48, 34, 92, 88, 43, 58

8 A naturalist measured the heights of 10 sequoia trees and got the following results, in feet: 260, 255, 275, 241, 238, 255, 221, 270, 265, and 250. Find the median height.

Mode

Another measure of central tendency used to describe a set of numbers is the *mode*. The **mode** is the number in the set that occurs most often. Recall the amounts raised by the friends selling cookies:

   $10, $10, $12, $15, $18, $19, $21, $23, $25, $28, $32, $35, $38, $39, $140

To find the mode, look for the number that appears most frequently.

EXAMPLE **Finding the Mode**

Find the mode of the amounts raised by the friends selling cookies.

• Arrange the numbers in order, or make a frequency table of the numbers.

   The numbers are arranged in order above.

• Select the number that appears most frequently.

   The most frequent amount raised is $10.

So, Cleve was using the mode when he said that the typical amount his friends raised selling cookies is $10.

A group of numbers may have no mode or more than one mode. Data that have two modes are called *bimodal*.

 Check It Out
 Find the mode.

   9 21, 23, 23, 29, 27, 22, 27, 27, 24, 24

   10 3.8, 4.2, 4.2, 4.7, 4.2, 8.1, 1.5, 6.4, 6.4

   11 3, 9, 11, 11, 9, 3, 11, 4, 8

   12 Soda in eight vending machines was selling for $0.75, $1, $1.50, $0.75, $0.95, $1, $0.75, and $1.25.

Range

The **range** tells how far apart the greatest and least numbers are in a set. Consider the highest points on the seven continents:

Continent	Highest Elevation
Africa	19,340 ft
Antarctica	16,864 ft
Asia	29,028 ft
Australia	7,310 ft
Europe	18,510 ft
North America	20,320 ft
South America	22,834 ft

To find the range, you must subtract the least altitude from the greatest.

EXAMPLE Finding the Range

Find the range of the highest elevations on the seven continents.

• Find the greatest and least values.

The greatest value is 29,028 and the least value is 7,310.

• Subtract.

$29,028 - 7,310 = 21,718$

The range is 21,718 ft.

 Check It Out

Find the range.

⑬ 250, 300, 925, 500, 15, 600

⑭ 3.2, 2.8, 6.1, 0.4

⑮ 48°, 39°, 14°, 26°, 45°, 80°

⑯ The following number of students stayed for after-school activities one week: 28, 32, 33, 21, 18.

4·4 Exercises

Find the mean, median, mode, and range. Round to the nearest tenth.

1. 3, 3, 4, 6, 6, 7, 8, 10, 10, 10
2. 20, 20, 20, 20, 20, 20, 20
3. 12, 9, 8, 15, 15, 13, 15, 12, 12, 10
4. 76, 84, 88, 84, 86, 80, 92, 88, 84, 80, 78, 90
5. Are any of the sets of data in Exercises 1–4 bimodal? Explain.

6. The highest point in Arizona is Mount Humphreys, at 12,633 feet, and the lowest point, 70 feet, is on the Colorado River. What is the range in elevations?

7. Moises had scores of 83, 76, 92, 76, and 93 on his history tests. Which of the mean, median, or mode do you think he should use to describe the test scores?

8. Does the mean have to be a member of the set of data?

9. The following numbers represent winning scores in baseball by the Treetown Tigers:

   3 2 4 2 30 2 1 2 4 7 2 1

   Find the mean, median, and mode of the scores. Round to the nearest tenth. Which measure best represents the data? Explain.

10. Are you using the mean, median, or mode when you say that most of the runners finished the race in 5 minutes?

4·5 Probability

If you and a friend want to decide who goes first in a game, you might flip a coin. You and your friend have an equal chance of winning the toss. The **probability** of an event is a number from 0 to 1 that measures the chance that an event will occur.

Simple Events

The **outcome** is any one of the possible results of an action. A **simple event** is one outcome or a collection of outcomes. When you and your friend flip a coin to see who goes first in a game, getting heads is a simple event. The chance of that event occurring is its probability.

If all outcomes are equally likely, the probability of a simple event is a ratio that compares the number of favorable events outcomes to the number of possible outcomes.

$$P(\text{event}) = \frac{\text{number of favorable events outcomes}}{\text{total number of possible outcomes}}$$

The coin that you flip to decide who goes first has two sides: a head and a tail.

$$P(\text{head}) = \frac{1 \text{ favorable outcome}}{2 \text{ possible outcomes}} = \frac{1}{2}$$

Suppose that you are playing a game with number cubes that are numbered 1–6. There are six possible outcomes when you roll the number cube. These outcomes are equally likely, so the probability of rolling a 6 is $\frac{1}{6}$, or $P = \frac{1}{6}$.

 Check It Out

Use the number cube above to find each probability.

1. $P(\text{odd number})$
2. $P(\text{even number})$
3. $P(\text{prime number})$
4. $P(\text{number greater than 4})$

The spinner is $\frac{3}{4}$ red and $\frac{1}{4}$ yellow. The probability of spinning red is $\frac{3}{4}$. The probability of spinning yellow is $\frac{1}{4}$, or an unlikely event. The probability of spinning blue is impossible because there is no blue section on the spinner.

EXAMPLE Finding the Probability of a Simple Event

What is the probability of the spinner below landing on green? Are all events equally likely?

• Compare the probabilities of the spinner landing on each color.
• Use the formula to find the probability.

$$P(\text{event}) = \frac{\text{number of favorable events outcomes}}{\text{number of possible outcomes}}$$

$$P = \frac{1}{4}$$

The spinner is equally divided into 4 equal sections, so the probability of landing on green is equally likely to that of landing on yellow, red, or blue.

 Check It Out

Use the spinner to find the probability of an event.

5 $P(\text{blue})$

6 $P(\text{red})$

7 $P(\text{black})$

Expressing Probabilities

You can express a probability as a fraction, as shown before. But just as you can write a fraction as a decimal, ratio, or percent, you can also write a probability in any of those forms.

Fraction	Decimal	Ratio	Percent
$\frac{1}{2}$	0.5	1:2	50%

 Check It Out

Express each of the following probabilities as a fraction, decimal, ratio, and percent.

8 the probability of drawing a vowel from cards containing the letters of the name *Washington*

9 the probability of getting a red marble when you are drawing a marble from a bag containing two red marbles and six blue ones

---

EXAMPLE Expressing Probability

What is the probability of rolling an even number on a standard number cube? Express probability as a fraction, decimal, ratio, and percent.

- Use the formula to find the probability.

$$P(\text{even numbers}) = \frac{\text{even numbers possible}}{\text{total numbers possible}}$$
$$= \frac{3}{6} \text{ or } \frac{1}{2}$$

The probability of rolling an even number is $\frac{1}{2}$, 0.5, 1:2, or 50%.

Outcome Grids

One way to show the outcomes in an experiment is to use an **outcome grid**.

EXAMPLE Making Outcome Grids

Make an outcome grid to show the results of tossing two coins.

• List the outcomes of tossing a coin in the top row and in the left column.

		2nd toss	
		Head	**Tail**
1st toss	**Head**		
	Tail		

• Fill in the outcomes.

		2nd toss	
		Head	**Tail**
1st toss	**Head**	H, H	H, T
	Tail	T, H	T, T

After you have completed the outcome grid, it is easy to count target outcomes and determine probabilities.

Check It Out

Use the spinner to answer Exercises 10 and 11.

10 Make an outcome grid to show the two-digit outcomes when spinning the spinner twice. The first spin produces the first digit. The second spin produces the second digit.

11 What is the probability of getting a number divisible by 11 when you spin the spinner twice?

Probability Line

You know that the probability of an event is a number from 0 to 1. One way to show probabilities and how they relate to each other is to use a **probability line**. The following probability line shows the possible ranges of probability values.

Impossible   Equally Likely   Certain
Event   Events   Event

$$0 \qquad\qquad \frac{1}{2} \qquad\qquad 1$$

The line shows that events that are certain to happen have a probability of 1. Such an event is the probability of getting a number between 0 and 7 when a standard number cube is rolled.

An event that cannot happen has a probability of zero. The probability of getting an 8 when spinning a spinner that shows 0, 2, and 4 is 0. Events that are equally likely, such as getting a head or a tail when you toss a coin, have a probability of $\frac{1}{2}$.

EXAMPLE   Showing Probability on a Probability Line

If you toss two coins, you can use a number line to show the probabilities of getting two tails and of getting a tail and a head.

• Draw a number line, and label it from 0 to 1.

$$0 \qquad\qquad \frac{1}{2} \qquad\qquad 1$$

• Calculate the probabilities of the given events, and show them on the probability line.

From the outcome grid on page 210, you can see that you get two tails once and a head and a tail twice. $P(2 \text{ tails})$ $= \frac{1}{4}$ and $P(\text{head and tail}) = \frac{2}{4} = \frac{1}{2}$. The probabilities are shown on the following probability line.

$$\boxed{P(2 \text{ tails})}\;\boxed{P(\text{head and tail})}$$

$$0 \qquad \frac{1}{4} \qquad \frac{1}{2} \qquad\qquad 1$$

Check It Out

Draw a probability line. Then plot the following:

12 the probability of drawing a red marble from a bag of red marbles

13 the probability of rolling a 5 or a 6 on one roll of a die

14 the probability of getting both heads or both tails when flipping a coin twice

15 the probability of drawing a white marble from a bag of red marbles

Tree Diagrams

You can use a tree diagram to count outcomes. For example, suppose that you have two **spinners**—one with the numbers 1 and 2 and another that is half-red and half-green. You want to find out all the possible outcomes if you spin the spinners. You can make a **tree diagram**.

To make a tree diagram, you list what can happen with the first spinner.

First spinner

1

2

Then you list what can happen with the second spinner.

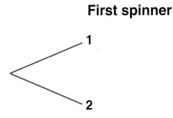

First spinner	Second spinner	List of possibilities
1	red	1, red
	green	1, green
2	red	2, red
	green	2, green

After listing the possibilities, you can count to see how many there are. In this case, there are four possible results.

To find the number of results or possibilities, you multiply the number of choices at each step:

$$2 \times 2 = 4$$

Check It Out

Draw a tree diagram. Check by multiplying.

16 Salvador has red, blue, and green shirts and black, white, and gray slacks. How many possible outfits does he have?

17 Balls come in red, blue, and orange and in large and small sizes. If a store owner has a bin for each type of ball, how many bins are there?

18 If you toss three coins, how many possible ways can they land?

19 How many possible ways can Jamie walk from school to home if she goes to the library on the way home?

Permutations

The tree diagram shows ways that things can be arranged, or listed. A listing in which the order is important is called a **permutation**. Suppose that you want to take a picture of the dogs Spot, Brownie, and Topsy. They could be arranged in any of several ways. You can use a tree diagram to show the ways.

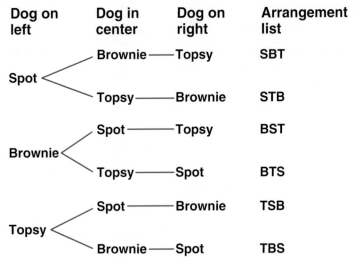

Dog on left	Dog in center	Dog on right	Arrangement list
Spot	Brownie — Topsy		SBT
	Topsy — Brownie		STB
Brownie	Spot — Topsy		BST
	Topsy — Spot		BTS
Topsy	Spot — Brownie		TSB
	Brownie — Spot		TBS

There are 3 choices for the first dog, 2 choices for the second, and 1 choice for the third, so the total number of permutations is $3 \times 2 \times 1 = 6$. Remember that Spot, Brownie, Topsy is a different permutation from Topsy, Brownie, Spot.

$P(3, 3)$ represents the number of permutations of 3 things taken 3 at a time. So, $P(3, 3) = 6$.

EXAMPLE Finding Permutations

Find $P(5, 3)$.

- Determine how many choices there are for each place.

  There are 5 choices for the first place, 4 for the second, and 3 for the third.

- Find the product.

  $5 \times 4 \times 3 = 60$

So, $P(5, 3) = 120$.

Factorial Notation

You saw that to find the number of permutations of 3 things, you found the product $3 \times 2 \times 1$. The product $3 \times 2 \times 1$ is called 3 **factorial**. The shorthand notation for factorial is an exclamation point. So, $3! = 3 \times 2 \times 1$.

 Check It Out

Solve.

20 $P(4, 3)$

21 $P(6, 6)$

22 The dog show has four finalists. In how many ways can four prizes be awarded?

23 One person from the 10-member swim team will represent the team at the city finals, and another member will represent the team at the state finals. In how many ways can you choose the two people?

Find each value. Use a calculator if needed.

24 8!

25 $6! \div 4!$

Combinations

When you choose two delegates from a class of 35 to attend a science fair, the order is not important. That is, choosing Norma and Miguel is the same as choosing Miguel and Norma when you are choosing two delegates.

You can use the number of permutations to find the number of **combinations**. Suppose that you and four friends (Gabe, Rita, Itay, and Kei) are going to the amusement park. There is room in your car for three of them. What different combinations of friends can ride with you?

Friend one	Friend two	Friend three	List
Gabe	Rita	Itay	GRI
		Kei	GRK
	Itay	Rita	GIR
		Kei	GIK
	Kei	Rita	GKR
		Itay	GKI
Rita	Gabe	Itay	RGI
		Kei	RGK
	Itay	Gabe	RIG
		Kei	RIK
	Kei	Gabe	RKG
		Itay	RKI
Itay	Gabe	Rita	IGR
		Kei	IGK
	Rita	Gabe	IRG
		Kei	IRK
	Kei	Gabe	IKG
		Rita	IKR
Kei	Gabe	Rita	KGR
		Itay	KGI
	Rita	Gabe	KRG
		Itay	KRI
	Itay	Gabe	KIG
		Rita	KIR

Because order does not matter, there are duplicates.

For example, GRI, GIR, RGI, RIG, IGR, and IRG all represent the same combination.

You have four choices for the first person, three choices for the second, and two choices for the third ($4 \times 3 \times 2 = 24$). But the order does not matter, so some combinations were counted too often! You need to divide by the number of different ways three objects can be arranged (3!).

$$\frac{4 \times 3 \times 2}{3 \times 2 \times 1} = 4$$

So there are four different groups of friends that can go in your car.

$C(4, 3)$ represents the number of combinations of four objects taken three at a time. Therefore $C(4, 3) = 4$.

EXAMPLE Finding Combinations

Find $C(5, 2)$.

- Find the permutations of the number of choices taken a number of times.

    $P(5, 2)$
    $P(5, 2) = 5 \times 4 = 20$

- Divide the permutations by the number of ways they can be arranged.

    Two times can be arranged 2! ways.
    $20 \div 2! = 20 \div 2 = 10$

So, $C(5, 2) = 10$.

 Check It Out

Find each value.

26 $C(8, 6)$

27 $C(12, 3)$

28 How many different combinations of four books can you choose from nine books?

29 Are there more combinations or permutations of three cats from a total of 12? Explain.

What are your initials? Do you have anything with your monogram on it? A *monogram* is a design that is made up of one or more letters, usually the initials of a name. Monograms often appear on stationery, towels, shirts, or jewelry.

How many different three-letter monograms can you make with the letters of the alphabet? Use a calculator to compute the total number. Remember to allow for repeat letters in the combination. See **Hot**Solutions for the answers.

Experimental Probability

The probability of an event is a number from 0 to 1. One way to find the probability of an event is to conduct an experiment.

Suppose you want to know the probability of your winning a game of tennis. You play 12 games and win 8 of them. To find the probability of winning, you can compare the number of games you win to the number of games you play. In this case, the **experimental probability** that you will win is $\frac{8}{12}$, or $\frac{2}{3}$.

EXAMPLE Determining Experimental Probability

Find the experimental probability of finding your friend at the library when you go.

- Conduct an experiment. Record the number of trials and the result of each trial.

   Go to the library 10 different times, and see whether your friend is there. Suppose your friend is there 4 times.

- Compare the number of occurrences of one set of results with the number of trials. That is the probability for that set of results.

   Compare the number of times your friend was there with the total number of times you went to the library.

The experimental probability of finding your friend at the library for this test is $\frac{4}{10}$, or $\frac{2}{5}$.

Check It Out

A marble is drawn from a bag of 20 marbles. Each time, the marble was returned before the next one was drawn. The results are shown on the circle graph.

Marbles Drawn

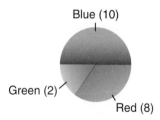

Blue (10)

Green (2)

Red (8)

30 Find the experimental probability of getting a red marble.

31 Find the experimental probability of getting a green marble.

32 Place 50 counters of different colors in a bag. Find the experimental probability that you will choose a counter of one color. Compare your answers with others' answers.

Theoretical Probability

You can find the **theoretical probability** if you consider the outcomes of the experiment.

EXAMPLE Determining Theoretical Probability

Find the probability of rolling an even number when you roll a number cube containing the numbers 1–6.

- Determine the number of ways the event occurs.

    In this case, the event is getting an even number. There are three even numbers—2, 4, and 6—on the number cube.

- Determine the total number of outcomes. Use a list, multiply, or make a *tree diagram* (p. 212).

    There are six numbers on the cube.

- Use the formula:
$$P(\text{event}) = \frac{\text{Number of ways an event occurs}}{\text{Number of outcomes}}$$

- Find the probability of the target event.

    Find the probability of rolling an even number, represented by $P(\text{even})$.

    $P(\text{even}) = \frac{3}{6} = \frac{1}{2}$

The probability of rolling an even number is $\frac{1}{2}$.

Check It Out

Find each probability. Use the spinner for Exercises 33 and 34.

33 $P(\text{red})$

34 $P(\text{green})$

35 $P(\text{odd number})$ when tossing a 1–6 number cube

36 The letters of the word *mathematics* are written on identical slips of paper and placed in a bag. If you draw a slip at random, what is the probability that it will be an *m*?

Independent Events

If you roll a number cube and toss a coin, the result of one does not affect the other. These events are called **independent events**.

To find the probability of getting a 4 and then a tail, you can find the probability of each event and then multiply. The probability of getting a 4 on a roll of the number cube is $\frac{1}{6}$, and the probability of getting a tail is $\frac{1}{2}$. So the probability of getting a 4 and a tail is $\frac{1}{6} \times \frac{1}{2} = \frac{1}{12}$.

To find the probability of independent or dependent events:
- Find the probability of the first event.
- Find the probability of the second event.
- Find the product of the two probabilities.

EXAMPLE Find Probability of an Independent Event

Find the probability of spinning 6 on the spinner and rolling 3 on a number cube.

$\frac{1}{6}$ • Find the probability of the first event—spinning 6.

$\frac{3}{6}$ • Find the probability of the second event—rolling 3.

$\frac{1}{6} \times \frac{3}{6} = \frac{3}{12}$ • Find the product of the two probabilities.

$\frac{3}{12} = \frac{1}{4}$ • Simplify, if possible.

The probability of spinning 6 on the spinner and rolling 3 on the number cube is $\frac{1}{4}$.

Dependent Events

Suppose that you have seven yellow and three white tennis balls in a bag. The probability that you choose a white tennis ball at random is $\frac{3}{10}$. After you have taken a white tennis ball out, however, there are only nine balls left, two of which are white. So, the probability that a friend gets a white tennis ball after you have drawn one out is $\frac{2}{9}$. These events are called **dependent events** because the probability of one depends on the other.

In the case of dependent events, you multiply to find the probability of both events happening. So, the probability that your friend chooses a white tennis ball and you also choose one is $\frac{3}{10} \times \frac{2}{9} = \frac{1}{15}$.

EXAMPLE **Finding Probability of a Dependent Event**

There are 4 blueberry, 6 carrot, and 2 whole-wheat muffins in a bag. Teresa randomly selects 2 muffins without replacing the first muffin. Find the probability that she selects a carrot muffin and then a whole-wheat muffin.

$\frac{6}{12}$ • Find the probability of the first event.

$\frac{2}{11}$ • Find the probability of the second event.

$\frac{6}{12} \times \frac{2}{11} = \frac{12}{132}$ • Find the product of the two probabilities.

$\frac{12}{132} = \frac{1}{11}$ • Simply, if possible.

The probability of selecting a carrot muffin and then a whole-wheat muffin is $\frac{1}{11}$.

Check It Out

Solve.

37 Find the probability of rolling a 5 and an even number if you roll two number cubes. Are the events dependent or independent?

4·5 Exercises

**You roll a six-sided number cube numbered 1 through 6.
Find each probability as a fraction, decimal, ratio, and percent
in Exercises 1 and 2.**

1. $P(4$ or $5)$ 2. P(even number)

3. If you draw a card from a regular deck of 52 cards, what is
the probability of getting an ace? Is this experimental or
theoretical probability?

4. If you toss a thumbtack 48 times and it lands up 15 times,
what is the probability of its landing up again on the next toss?

5. Draw a probability line to show the probability of getting a
number less than 1 when you are rolling a six-sided number
cube numbered 1 through 6.

6. Make an outcome grid to show the outcomes of spinning two
spinners containing the numbers 1–5.

7. Find the probability of drawing a red ball and a white ball
from a bag of 16 red and 24 white balls if you replace the
balls between drawings.

8. Find the probability of drawing a red ball and a white ball
(Exercise 7) if you do not replace the balls between
drawings.

9. In which of Exercises 7 and 8 are the events independent?

Solve.

10. Six friends want to play enough games of checkers to make
sure that everyone plays everyone else. How many games will
they have to play?

11. Eight people swim in the 200 meter finals at an Olympic
trial. Medals are given for first, second, and third place.
How many ways are there to give the medals?

12. Determine whether each of the following is a permutation or
a combination.
A. choosing first, second, and third place at an oratory
contest among 15 people
B. choosing four delegates from a class of 20 to attend
Government Day

Data, Statistics, and Probability

What have you learned?

You can use the problems and the list of words that follow to see what you learned in this chapter. You can find out more about a particular problem or word by referring to the topic number (*for example*, Lesson 4·2).

Problem Set

1. Mr. Chan took a survey of his class. He gave each student a number and then put duplicate numbers in a paper bag. He drew ten numbers from the bag without peeking. He then surveyed those ten students. Was this a random sample?
(Lesson 4·1)

2. A survey asked, "Are you a responsible citizen who has registered to vote?" Rewrite the question so that it is not biased.
(Lesson 4·1)

3. On a circle graph, how many degrees must be in a sector to show 25%? (Lesson 4·2)

Use the bar graph to answer Exercises 4 and 5. (Lesson 4·2)

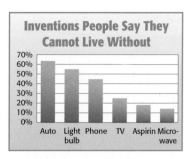

4. Which inventions did more than 50% of the people surveyed say they could not do without?

5. What percent of the people did not feel they could do without the microwave?

6. Describe the correlation shown below. (Lesson 4·3)

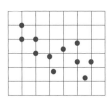

7. Find the mean, median, mode, and range of the numbers 18, 23, 18, 15, 20, and 17. (Lesson 4·4)

8. Of the mean, median, and mode, which must be a member of the set of data? (Lesson 4·4)

9. $C(9, 2) =$ _____ (Lesson 4·5)

10. $\dfrac{10!}{8!} =$ _____ (Lesson 4·5)

HotWords

Write definitions for the following words.

average (Lesson 4·4)

circle graph (Lesson 4·2)

combination (Lesson 4·5)

correlation (Lesson 4·3)

dependent events (Lesson 4·5)

double-bar graph (Lesson 4·2)

event (Lesson 4·5)

experimental probability (Lesson 4·5)

factorial (Lesson 4·5)

histogram (Lesson 4·2)

independent event (Lesson 4·5)

leaf (Lesson 4·2)

line graph (Lesson 4·2)

mean (Lesson 4·4)

median (Lesson 4·4)

mode (Lesson 4·4)

outcome (Lesson 4·5)

outcome grid (Lesson 4·5)

percent (Lesson 4·2)

permutation (Lesson 4·5)

population (Lesson 4·1)

probability (Lesson 4·5)

probability line (Lesson 4·5)

random sample (Lesson 4·1)

range (Lesson 4·4)

sample (Lesson 4·1)

scatter plot (Lesson 4·3)

simple event (Lesson 4·5)

spinner (Lesson 4·5)

stem (Lesson 4·2)

stem-and-leaf plot (Lesson 4·2)

survey (Lesson 4·1)

table (Lesson 4·1)

tally mark (Lesson 4·1)

theoretical probability (Lesson 4·5)

tree diagram (Lesson 4·5)

Hot Topic 5

Algebra

Problem Set

Write an equation for each sentence. (Lesson 5·1)

1. If 5 is subtracted from twice a number, the result is 3.
2. 6 times the sum of a number and 2 is 20.

Simplify each expression. (Lesson 5·2)

3. $7n + 6b - b - 4n$
4. $3(2n - 1) - (n + 4)$

5. Find the distance traveled by a jogger who jogs at 4 miles per hour for $2\frac{1}{2}$ hours. Use the formula $d = rt$. (Lesson 5·3)

Solve each equation. Check your solution. (Lesson 5·4)

6. $x + 3 = 9$
7. $\frac{y}{3} = -6$
8. $4x - 3 = 17$
9. $\frac{y}{5} - 3 = 4$
10. $7n - 8 = 4n + 7$
11. $y - 8 = 4y + 4$
12. $3(2n - 3) = 15$
13. $2(3x - 2) = 16$

Use a proportion to solve.

14. In a class, the ratio of boys to girls is $\frac{2}{3}$. If there are 10 boys in the class, how many girls are there? (Lesson 5·5)

Solve each inequality. Graph the solution. (Lesson 5·6)

15. $x + 4 < 3$ **16.** $3x + 2 \geq 11$

Locate each point on the coordinate plane and tell in which quadrant or on which axis it lies. (Lesson 5·7)

17. $A(-3, 4)$ **18.** $B(3, 0)$ **19.** $C(0, -2)$ **20.** $D(2, 3)$

21. Find the slope of the line that contains the points $(4, -2)$ and $(-2, 8)$.

Determine the slope and the *y*-intercept from the equation of each line. Graph the line. (Lesson 5·8)

22. $y = \frac{2}{3}x + 3$ **23.** $y = -2$ **24.** $y = 2x - 1$ **25.** $y = 3x$

HotWords

additive inverse (Lesson 5·4)

associative property (Lesson 5·2)

axes (Lesson 5·7)

Celsius (Lesson 5·3)

commutative property (Lesson 5·2)

cross product (Lesson 5·5)

difference (Lesson 5·1)

distributive property (Lesson 5·2)

equation (Lesson 5·1)

equivalent (Lesson 5·1)

equivalent expression (Lesson 5·2)

expression (Lesson 5·1)

Fahrenheit (Lesson 5·3)

formula (Lesson 5·3)

horizontal (Lesson 5·7)

inequality (Lesson 5·6)

like terms (Lesson 5·2)

order of operations (Lesson 5·3)

ordered pair (Lesson 5·7)

origin (Lesson 5·7)

perimeter (Lesson 5·3)

point (Lesson 5·7)

product (Lesson 5·1)

proportion (Lesson 5·5)

quadrant (Lesson 5·7)

quotient (Lesson 5·1)

rate (Lesson 5·5)

rate of change (Lesson 5·8)

ratio (Lesson 5·5)

slope (Lesson 5·8)

solution (Lesson 5·4)

sum (Lesson 5·1)

term (Lesson 5·1)

variable (Lesson 5·1)

vertical (Lesson 5·7)

x-axis (Lesson 5·7)

y-axis (Lesson 5·7)

y-intercept (Lesson 5·8)

5·1 Writing Expressions and Equations

Expressions

In mathematics, a placeholder, or variable, is used to represent a number. A **variable** is a symbol that represents an unknown quantity. Sometimes the first letters of important words make the meaning of the equation easier to remember.

An **expression** is a combination of variables or combinations of variables, numbers, and symbols that represent a mathematical relationship.

A **term** is a number, a variable, a product, or a quotient in an expression. A term is not a sum or a difference. In the expression $7x + 4b + 2$, there are three terms: $7x$, $4b$, and 2.

Expressions	
Word Expression	**Algebraic Expression**
A full box of markers $+$ 4 markers	$b + 4$
A full box of markers $-$ 4 markers	$b - 4$
Four full boxes of markers	$4b$ When you use variables in an expression, write multiplication without the $\times$.
Four full boxes of markers shared equally among 4 people	$\dfrac{b}{4}$ When you use variables in an expression, write division in fraction form.

Check It Out

Count the number of terms in each expression.

1 $3n + 8$

2 $4xyz$

3 $6ab - 2c - 5$

4 $\dfrac{3x}{y}$

Writing Expressions Involving Addition

Expressions are often interpretations of a written phrase. For example, the sentence "Jon can jump 3 feet higher than Ben can jump" can be written as an algebraic expression.

how high Ben can jump $= j$

Jon can jump higher $= + 3$ feet

The expression can be written $j + 3$.

Several words and phrases indicate addition. You might see the words *plus, more than, increased by, total,* or *altogether.* Another word that indicates addition is **sum**. The sum of two terms is the result of adding them together. For example, the sentence "Julie has increased her CD collection by 38" can be written as an algebraic expression. Let *c* represent the number of CDs that Julie had. The words *increased by* indicate addition. So, the expression is $c + 38$.

Here are some common addition phrases and the corresponding expressions.

Phrase	Expression
6 more than some number	$n + 6$
a number increased by 9	$x + 9$
4 plus some number	$4 + y$
the sum of a number and 7	$n + 7$

 Check It Out

Write an expression for each phrase.

5 a number added to 3

6 the sum of a number and 9

7 some number increased by 5

8 4 more than some number

Writing Expressions Involving Subtraction

Subtraction can be indicated by several words or phrases. You might see the words *minus, less than, less,* or *decreased by.* Another word that indicates subtraction is **difference**. The difference between two terms is the result of subtracting them. The sentence "Jodi jogged 9 meters less than she did yesterday" can be written $m - 9$. The variable m represents the unknown number of meters that Jodi jogged yesterday. Take away, or subtract, 9 meters from that number to find how many meters Jodi jogged today.

The order in which an expression is written is very important in subtraction. You need to know which term is being subtracted and which term is being subtracted from. For example, in the phrase "7 is less than a number," restate the expression, substituting the words "a number" with the term 10: What is 7 less than 10? Mathematically this is written as $10 - 7$, not $7 - 10$. The phrase is then translated to the algebraic expression $x - 7$, not $7 - x$.

Some common subtraction phrases and the corresponding expressions are listed below.

Phrase	Expression
9 less than some number	$n - 9$
a number decreased by 6	$x - 6$
10 minus some number	$10 - y$
the difference between a number and 7	$n - 7$

 Check It Out

Write an expression for each phrase.

9 a number subtracted from 10

10 the difference between a number and 3

11 some number decreased by 5

12 8 less than some number

Writing Expressions Involving Multiplication

Several words and phrases indicate multiplication. You might see the words *times, product, multiplied, of,* or *twice.* The sentence "The Rams scored twice as many points as the Pirates scored" can be written $2p$. The variable p represents the points scored by the Pirates, and 2 is used for twice as many, or 2 times. The operation between the variable and 2 is multiplication. The result of multiplying two terms is called the **product**.

Some common multiplication phrases and the corresponding expressions are listed below.

Phrase	Expression
3 times some number	$3a$
twice a number	$2x$
one-fifth of some number	$\frac{1}{5}y$
the product of a number and 6	$6n$

EXAMPLE **Writing Expressions Involving Multiplication**

Write an algebraic expression.

The garden produced 4 times as many beans as last year.

- Determine the operation required. The word *times* indicates multiplication.
- Let b represent the unknown number of beans.
- Write an algebraic expression.

$4b$

 Check It Out

Write an expression for each phrase.

13 a number multiplied by 6

14 the product of a number and 4

15 75% of some number

16 10 times some number

Writing Expressions Involving Division

The words *divide* and *ratio of* indicate the process of division. Another word that indicates division is **quotient**. The quotient of two terms is the result of one being divided by the other. The phrase "16 cookies divided by a number of friends" can be written as the expression $\frac{16}{c}$; the variable c represents the unknown number of friends. The words *divided by* indicate that the operation between the number and 16 is division.

Some common division phrases and the corresponding expressions are listed below.

Phrase	Expression
the quotient of 10 and some number	$\frac{10}{n}$
a number divided by 8	$\frac{x}{8}$
the ratio of 20 and some number	$\frac{20}{y}$
the quotient of a number and 3	$\frac{n}{3}$

 Check It Out

Write an expression for each phrase.

17 a number divided by 3

18 the quotient of 12 and a number

19 the ratio of 30 and some number

20 the quotient of some number and 7

Writing Expressions Involving Two Operations

To translate the phrase "2 added to the product of 3 and some number" to an expression, first realize that "2 added to" means "something" + 2. That "something" is "the product of 3 and some number," which is $3x$ because "product" indicates multiplication. So, the expression can be written as $3x + 2$.

Phrase	Expression	Think
2 less than the product of a number and 5	$5x - 2$	"2 less than" means "something" $-$ 2; "product" indicates multiplication.
5 times the sum of a number and 3	$5(x + 3)$	Write the sum inside parentheses so that the entire sum is multiplied by 5.
3 more than 7 times a number	$7x + 3$	"3 more than" means "something" $+$ 3; "times" indicates multiplication.

 Check It Out

Translate each phrase to an expression.

21 8 less than the product of 5 and a number

22 4 subtracted from the product of 2 and a number

23 twice the difference between a number and 10

Writing Equations

An expression is a phrase; an **equation** is a mathematical sentence. An equation indicates that two expressions are **equivalent**, or *equal*. The symbol used in an equation is the equals sign, "=."

To translate the sentence "3 less than the product of a number and 4 is 9" to an equation, first identify the words that indicate "equals." In this sentence, "equals" is indicated by "is the same as." In other sentences "equals" may be "is," "the result is," "you get," or just "equals."

After you have identified the =, you can then translate the phrase that comes before the = and write the expression on the left side. Then write the term = on the right side.

3 less than the product of a number and 4	is the same as	term
phrase before equals	=	number after equals
$4x - 3$	=	9

Check It Out

Write an equation for each sentence.

24 6 subtracted from 3 times a number is 6.

25 If 5 is added to the quotient of a number and 4, the result is 10.

26 2 less than 3 times a number is 25.

APPLICATION **Orphaned Whale Rescued**

On January 11, 1997, an orphaned baby gray whale arrived at an aquarium in California. Rescue workers named her J.J. She was three days old, weighed 1,600 pounds, and was desperately ill.

Soon her caretakers had her sucking from a tube attached to a thermos. By February 7, on a diet of whale milk formula, she weighed 2,378 pounds. J.J. was gaining 20 to 30 pounds a day!

An adult gray whale weighs approximately 35 tons, but the people at the aquarium knew that they could release J.J. when she had a solid layer of blubber—or when she weighed about 9,000 pounds.

Write an equation showing that J.J. is 2,378 pounds now and needs to gain 25 pounds per day for some number of days until she weighs 9,000 pounds. See **HotSolutions** for the answer.

WRITING EXPRESSIONS AND EQUATIONS

5·1

5·1 Exercises

Count the number of terms in each expression.

1. $3x + 5$ **2.** 7   **3.** $5x - 2$ **4.** $8n - 13$

Write an expression for each phrase.

5. 6 more than a number

6. a number added to 9

7. the sum of a number and 5

8. 4 less than a number

9. 10 decreased by some number

10. the difference between a number and 6

11. one fourth of some number

12. twice a number

13. the product of a number and 4

14. a number divided by 6

15. the ratio of 8 and some number

16. the quotient of a number and 3

17. 7 more than the product of a number and 4

18. 1 less than twice a number

Write an equation for each sentence.

19. 6 more than the quotient of a number and 4 is 8.

20. If 5 is subtracted from twice a number, the result is 11.

21. 4 times the sum of a number and 3 is 12.

Select the correct response.

22. Which of the following words is used to indicate multiplication?
   A. sum   **B.** difference   **C.** product   **D.** quotient

23. Which of the following does not indicate subtraction?
   A. less than   **B.** difference
   C. decreased by   **D.** ratio of

24. Which of the following shows "twice the sum of a number and 8"?
   A. $2(x + 8)$ **B.** $2x + 8$ **C.** $(x - 8)$ **D.** $2 + (x + 8)$

Terms

As you may remember, terms can be numbers, variables, or numbers and variables combined by multiplication or division. Some examples of terms are listed below.

$$n \qquad 13 \qquad 3x \qquad x^3$$

Compare the terms 13 and $3x$. The value of $3x$ will change as the value of x changes. If $x = 2$, then $3x = 3(2) = 6$, and if $x = 3$, then $3x = 3(3) = 9$. Notice, though, that the value of 13 never changes—it remains constant. When a term contains a number only, it is called a *constant* term.

 Check It Out

Decide whether each term is a constant term.

1 $5x$ **2** 8

3 $4(n + 2)$ **4** 3

The Commutative Property of Addition and Multiplication

The **Commutative Property** of Addition states that the order of terms being added may be switched without changing the result: $6 + 8 = 8 + 6$ and $b + 9 = 9 + b$. The Commutative Property of Multiplication states that the order of terms being multiplied may be switched without changing the result: $6(8) = 8(6)$ and $b \times 9 = 9b$.

The Commutative Property does not hold for subtraction or division. The order of the terms does affect the result: $9 - 3 = 6$, but $3 - 9 = -6$; $12 \div 4 = 3$, but $4 \div 12 = \frac{1}{3}$.

Check It Out

Rewrite each expression, using the Commutative Property of Addition or Multiplication.

5 $2x + 3$

6 $n \times 5$

7 $6 + 3y$

8 5×6

The Associative Property of Addition and Multiplication

The **Associative Property** of Addition states that the grouping of terms being added does not affect the result:
$(6 + 4) + 7 = 6 + (4 + 7)$ and $(x + 3) + 5 = x + (3 + 5)$.
The Associative Property of Multiplication states that the grouping of terms being multiplied does not affect the result:
$(8 \times 3) \times 5 = 8 \times (3 \times 5)$ and $7 \times 4b = (7 \times 4)b$.

The Associative Property does not hold for subtraction or division. The grouping of the numbers does affect the result:
$(10 - 8) - 6 = -2$, but $10 - (8 - 6) = 8$; $(20 \div 2) \div 2 = 5$, but $20 \div (2 \div 2) = 20$.

Check It Out

Rewrite each expression, using the Associative Property of Addition or Multiplication.

9 $(3 + 7) + 10$

10 $(4 \times 2) \times 7$

11 $(4x + 3y) + 5$

12 $4 \times 5n$

The Distributive Property

The **Distributive Property** of Addition and Multiplication states that multiplying a sum by a number is the same as multiplying each addend by that number and then adding the products. So, $4(6 + 5) = (4 \times 6) + (4 \times 5)$.

How would you multiply 5×49 in your head? You might think, $250 - 5 = 245$. If you did, you have used the Distributive Property.

$5(50 - 1)$

$= 5 \times 50 - 5 \times 1$ • Distribute the factor of 5 to each term inside the parentheses.

$= 250 - 5$ • Simplify, using order of operations.

$= 245$

The Distributive Property does not hold for division.
$6 \div (2 + 1) \neq (6 \div 2) + (6 \div 1)$

Check It Out

Use the Distributive Property to find each product.

13 3×48

14 6×99

15 5×85

16 7×199

Equivalent Expressions

The Distributive Property can be used to write an **equivalent expression** with two terms. Equivalent expressions are two different ways of writing one expression.

EXAMPLE **Writing an Equivalent Expression**

Write an equivalent expression for $2(3x - 1)$.

$2(3x - 1)$ • Distribute the factor to each term inside the parentheses.

$= 2 \times 3x - 2 \times 1$ • Simplify.

$= 6x - 2$ • Write the equivalent expressions.

$2(3x - 1) = 6x - 2$

Distributing When the Factor Is Negative

The Distributive Property is applied in the same way if the factor to be distributed is negative.

EXAMPLE **Distributing When the Factor Is Negative**

Write an equivalent expression for $-3(5x - 1)$.

$-3(5x - 1)$ • Distribute the factor to each term inside the parentheses.

$= -3 \times 5x - (-3) \times 1$ • Simplify. (Remember: $(-3) \times 1 = -3$ and $-(-3) = +3$.)

$= -15x + 3$ • Write the equivalent expressions.

$-3(5x - 1) = -15x + 3$

Check It Out

Write an equivalent expression.

17 $2(5x + 2)$

18 $6(3n - 2)$

19 $-1(6y - 4)$

20 $-2(-4x + 5)$

The Distributive Property with Common Factors

For the expression $8n + 16$, you can use the Distributive Property to write an equivalent expression. Recognize that each of the two terms has a factor of 4.

Rewrite the expression as $4 \times 2n + 4 \times 4$. Then write the common factor 4 in front of the parentheses and the remaining factors inside the parentheses: $4(2n + 4)$. You have used the Distributive Property to *factor out the common factor.*

EXAMPLE	**Factoring Out the Common Factor**

Factor out the common factor from the expression $14n - 35$.

$14n - 35$ • Find a common factor.

$7 \times 2 - 7 \times 5$ • Rewrite the expression.

$7 \times (2n - 5)$ • Use the Distributive Property.

So, $14n - 35 = 7(2n - 5)$.

When you factor, always make sure that you factor out the greatest common factor.

 Check It Out

Factor out the greatest common factor in each expression.

㉑ $7x + 21$

㉒ $12n - 9$

㉓ $10c + 30$

㉔ $10b + 25$

Like Terms

Like terms are terms that contain the same variable with the same exponent. Constant terms are like terms because they have no variables. Here are some examples of like terms:

Like Terms	Reason
$6x$ and $9x$	Both contain the same variables.
2 and 7	Both are constant terms.
$3n^2$ and $9n^2$	Both contain the same variable with the same exponent.

Some examples of terms that are not like terms are listed.

Not Like Terms	Reason
$2x$ and $7y$	Variables are different.
$5n$ and 10	One term has a variable; the other is constant.
$4x^2$ and $12x$	The variables are the same, but the exponents are different.

Like terms may be combined into one term by adding or subtracting. Consider the expression $3x + 4x$. Notice that the two terms have a common factor, x. Use the Distributive Property to write $x(3 + 4)$. This simplifies to $7x$, so $3x + 4x = 7x$.

EXAMPLE Combining Like Terms

Simplify $4y - 8y$.

$y(4 - 8)$
- Recognize that the variable is a common factor. Rewrite the expression, using the distributive property.

$y(-4)$
- Simplify.

$-4y$
- Use the Commutative Property of Multiplication.

So, $4y - 8y$ simplified is $-4y$.

Check It Out

Combine like terms.

25 $4x + 9x$

26 $10y - 6y$

27 $5n + 4n + n$

28 $3a - 7a$

Simplifying Expressions

Expressions are simplified when all of the like terms have been combined. Terms that are not like terms cannot be combined. In the expression $3x - 5y + 6x$, there are three terms. Two of them are like terms, $3x$ and $6x$, which combine to $9x$. The expression can be written as $9x - 5y$, which is simplified because the two terms are not like terms.

EXAMPLE Simplifying Expressions

Simplify the expression $4(2n - 3) - 10n + 17$.

$4(2n - 3) - 10n + 17$

• Combine like terms, if any.

• Use the Distributive Property.

$= 4 \times 2n - 4 \times 3 - 10n + 17$ • Simplify.

$= 8n - 12 - 10n + 17$ • Combine like terms.

$= -2n + 5$

• If the remaining terms are not like terms, the expression is simplified.

So, $4(2n - 3) - 10n + 17$ simplified is $-2n + 5$.

Check It Out

Simplify each expression.

29 $7y + 5z - 2y + z$

30 $x + 3(2x - 5)$

31 $8a + 6a - 4(2 \times 3)$

32 $12n + 8 + 2(4n - 6)$

5·2 Exercises

Decide whether each term is a constant term. Write *yes* or *no*.

1. $8n$

2. -7

Rewrite each expression, using the Commutative Property of Addition or Multiplication.

3. $4 + 6$

4. $n \times 7$

5. $4x + 5$

Rewrite each expression, using the Associative Property of Addition or Multiplication.

6. $4 + (5 + 9)$

7. $(7 \times 5) \times 2$

8. $2 \times 5n$

Use the Distributive Property to find each product.

9. 4×99

10. 6×104

Write an equivalent expression.

11. $3(5x + 4)$

12. $-5(2n + 6)$

13. $10(3a - 7)$

14. $-(-4y - 6)$

Factor out the greatest common factor in each expression.

15. $8x + 16$

16. $12n - 4$

17. $20a - 30$

Combine like terms.

18. $10x - 7x$

19. $5n + 6n - n$

20. $3y - 8y$

Simplify each expression.

21. $8n + b - 2n - 4b$

22. $5x + 2(3x - 5) + 2$

23. $-2(-5n - 3) - (n + 2)$

24. Which property is illustrated by $5(2x + 1) = 10x + 5$?
    A. Commutative Property of Multiplication
    B. Distributive Property
    C. Associative Property of Multiplication
    D. The example does not illustrate a property.

25. Which of the following shows the expression $24x - 36$ with the greatest common factor factored out?
    A. $2(12x - 18)$ B. $3(8x - 12)$
    C. $6(4x - 6)$ D. $12(2x - 3)$

5·3 Evaluating Expressions and Formulas

Evaluating Expressions

After an expression has been written, you can *evaluate* it for different values of the variable. To evaluate $2x - 1$ for $x = 4$, *substitute* 4 in place of the x: $2(4) - 1$. Use **order of operations** to evaluate: multiply first, and then subtract. So, $2(4) - 1 = 8 - 1 = 7$.

EXAMPLE **Evaluating an Expression**

Evaluate $2x^2 - 4x + 3$ for $x = 3$.

$2x^2 - 4x + 3$, when $x = 3$
- Substitute the numeric value for variable.

$= 2(3^2) - 4 \times 3 + 3$
- Use order of operations to simplify. Simplify within parentheses, and then evaluate.

$= 2 \times 9 - 12 + 3$
- Multiply and divide in order from left to right.

$18 - 12 + 3 = 9$
- Add and subtract in order from left to right.

When $x = 3$, then $2x^2 - 4x + 3 = 9$.

Check It Out

Evaluate each expression for the given value.

1. $6x - 10$, for $x = 4$
2. $4a + 5 + a^2$, for $a = -3$
3. $\frac{n}{4} + 2n - 3$, for $n = 8$
4. $2(y^2 - y - 2) + 2y$, for $y = 3$

Evaluating Formulas

The Formula for Converting Fahrenheit Temperature to Celsius

Thermometers measure temperature in various units. Many parts of the world use the metric unit, the **Celsius**. In the United States, people use the **Fahrenheit** scale for weather and everyday purposes. One unit Fahrenheit equals $\frac{9}{5}$ of one Celsius unit. The formula $C = \frac{5}{9}(F - 32)$ can be used to convert Fahrenheit to Celsius.

EXAMPLE Converting Fahrenheit Temperature to Celsius

The highest temperature in Ohio last summer was 98°F. Use the formula $C = \frac{5}{9}(F - 32)$ to find this temperature in Celsius.

$98 - 32 = 66$
- Subtract 32 from the Fahrenheit temperature.

$5 \div 9 = 0.5555555555$
- Divide 5 by 9.

$0.5555555555 \times 66 = 36.7$
- Multiply the repeating decimal by the difference in temperature.

So, 98°F equals 36.7°C.

 Check It Out

Find each temperature in Celsius.

5 68°F

6 −5°F

7 32°F

EXAMPLE Converting Celsius Temperature to Fahrenheit

Use the formula $F = \frac{9}{5}C + 32$ to convert 37°C to Fahrenheit.

$9 \div 5 = 1.8$
- Divide 9 by 5.

$1.8 \times 37 = 66.6$
- Multiply 1.8 by the temperature.

$66.6 + 32 = 98.6$
- Add 32 to the product.

So, 37°C equals 98.6°F.

Check It Out

Convert each Celsius temperature to degrees Fahrenheit.

8 43.3°C

9 12.7°C

10 96°C

11 31°C

The Formula for Distance Traveled

The distance traveled by a person, vehicle, or object depends on its rate and the amount of time. The formula $d = rt$ can be used to find the distance traveled, d, if the rate, r, and the amount of time, t, are known.

EXAMPLE Finding the Distance Traveled

Find the distance traveled by a runner who averages 4 miles per hour for $3\frac{1}{2}$ hours.

$d = 4 \times 3\frac{1}{2}$ • Substitute values into the distance formula ($d = rt$).

$= 14$ mi   • Multiply.

The runner ran 14 miles.

 Check It Out

Find the distance traveled.

12 A person rides 15 miles per hour for 2 hours.

13 A plane flies 700 kilometers per hour for $3\frac{1}{4}$ hours.

14 A person drives a car 55 miles per hour for 6 hours.

15 A snail moves 2.4 feet per hour for 5 hours.

5·3 Exercises

Evaluate each expression for the given value.

1. $5x - 11$, for $x = 6$

2. $3a^2 + 7 - 2a$, for $a = 4$

3. $\frac{n}{6} - 2n + 8$, for $n = -6$

4. $3(2y - 1) - 4y + 6$, for $y = 4$

Use the formula $C = \frac{5}{9}(F - 32)$ to convert degrees Fahrenheit to Celsius.

5. $76°F$

6. $18°F$

Use the formula $F = \frac{9}{5}C + 32$ to convert degrees Celsius to Fahrenheit.

7. $3°C$

8. $69°C$

Use the formula $d = rt$ to answer Exercises 9–11.

9. Find the distance traveled by a walker who walks at 4 miles per hour for $1\frac{1}{2}$ hours.

10. A race car driver averaged 140 miles per hour. If the driver completed the race in $2\frac{1}{2}$ hours, how many miles was the race?

11. The speed of light is approximately 186,000 miles per second. About how far does light travel in 5 seconds?

5.4 Solving Linear Equations

True or False Equations

The equation $3 + 4 = 7$ represents a true statement. The equation $1 + 4 = 7$ represents a false statement. What about the equation $x + 4 = 7$? You cannot determine whether it is true or false until a value for x is known.

EXAMPLE Determine Whether the Equation Is True or False

Determine whether the equation $3x - 2 = 13$ is true or false for $x = 1$, $x = 3$, and $x = 5$.

$3x - 2 = 13$	$3x - 2 = 13$	$3x - 2 = 13$
$3(1) - 2 \overset{?}{=} 13$	$3(3) - 2 \overset{?}{=} 13$	$3(5) - 2 \overset{?}{=} 13$
$3 - 2 \overset{?}{=} 13$	$9 - 2 \overset{?}{=} 13$	$15 - 2 \overset{?}{=} 13$
$1 \neq 13$	$7 \neq 13$	$13 = 13$
False	False	True

 Check It Out

Determine whether each equation is *true* or *false* for $x = 2$, $x = 5$, and $x = 8$.

1. $6x - 3 = 9$
2. $2x + 3 = 13$
3. $5x - 7 = 18$
4. $3x - 8 = 16$

The Solution of an Equation

If you look back over the past equations, you will notice that each equation had only one value for the variable that made the equation true. This value is called the **solution** of the equation. If you were to try other values for x in the equations, all would give false statements.

EXAMPLE Determining a Solution

Determine whether 6 is the solution of the equation
$4x - 5 = 2x + 6$.

$4x - 5 = 2x + 6$	• Substitute possible solution for x.
$4(6) - 5 \overset{?}{=} 2(6) + 6$	• Simplify, using order of operations.
$24 - 5 \overset{?}{=} 12 + 6$	
$19 \neq 18$	

Because the statement is false, 6 is not the solution.

Check It Out

Determine whether the given value is the solution of the equation.

5 $6; 3x - 5 = 13$

6 $7; 2n + 5 = 3n - 5$

7 $4; 7(y - 2) = 10$

8 $1; 5x + 4 = 12x - 3$

Equivalent Equations

An *equivalent equation* can be obtained from an existing equation in one of four ways. Apply the same process to both sides of the equation to keep it balanced, or equal.

- Add the same term to both sides of the equation.
- Subtract the same term from both sides.
- Multiply by the same term on both sides.
- Divide by the same term on both sides.

Four equations equivalent to $x = 8$ are shown below.

Operation	Equation Equivalent to $x = 8$
Add 4 to both sides.	$x + 4 = 12$
Subtract 4 from both sides.	$x - 4 = 4$
Multiply by 4 on both sides.	$4x = 32$
Divide by 4 on both sides.	$\frac{x}{4} = 2$

 Check It Out

Write equations equivalent to $x = 12$.

9 Add 3 to both sides.

10 Subtract 3 from both sides.

11 Multiply by 3 on both sides.

12 Divide by 3 on both sides.

Additive Inverses

Two terms are **additive inverses** if their sum is 0. Some examples are -3 and 3, $5x$ and $-5x$, and $12y$ and $-12y$. The additive inverse of 7 is -7 because $7 + (-7) = 0$, and the additive inverse of $-8n$ is $8n$ because $-8n + 8n = 0$.

 Check It Out

Give the additive inverse of each term.

13 4 **14** $-x$ **15** -35 **16** $10y$

Solving Addition and Subtraction Equations

You can use equivalent equations to *solve* an equation. The solution is obtained when the variable is by itself on one side of the equation. The objective, then, is to use equivalent equations to isolate the variable on one side of the equation.

Consider the equation $x + 7 = 15$. For it to be considered solved, the x has to be on a side by itself. How can you get rid of the $+7$ that is also on that side? Remember that a term and its additive inverse add up to 0. The additive inverse of 7 is -7. To write an equivalent equation, subtract 7 from both sides.

APPLICATION Predicting Life Expectancy

How much longer are people expected to live in the future? Consider these statistics.

Year	Life Expectancy
1900	47.3
1920	54.1
1940	62.9
1960	69.7
1980	73.7
1990	75.4
2000	77.1

This life expectancy data can be roughly described by the equation $y = 0.3x - 526$.

In this equation, y represents life expectancy and x represents the year.

Why is this an unreasonable way to describe the data? See **HotSolutions** for the answer.

EXAMPLE Solving Equations Using Subtraction

Use additive inverses to solve $x + 7 = 15$.

$x + 7 - 7 = 15 - 7$	• Subtract 7 from both sides.
$x = 8$	• Simplify.
	Check the solution.
$8 + 7 \stackrel{?}{=} 15$	• Substitute the possible solution for x.
$15 = 15$	• Because this is a true statement, 8 is the correct solution.

So, in $x + 7 = 15$, $x = 8$.

EXAMPLE Solving Equations Using Addition

Use additive inverses to solve $n - 3 = 10$.

$n - 3 + 3 = 10 + 3$	• Write an equivalent equation with n on a side by itself. Notice that there is a $- 3$ on the same side as the variable. Its additive inverse is $+ 3$, so add 3 to both sides of the equation.
$n = 13$	• Simplify.
	Check the solution.
$(13) - 3 \stackrel{?}{=} 10$	• Substitute the possible solution for n.
$10 = 10$	• Because this is a true statement, 13 is the solution.

So, in $n - 3 = 10$, $n = 13$.

Check It Out

Solve each equation. Check your solution.

17 $x + 4 = 11$

18 $n - 5 = 8$

19 $y + 8 = 2$

20 $b - 5 = 1$

Solving Multiplication and Division Equations

Consider the equation $3x = 15$. Notice that no term is being added to or subtracted from the term with the variable. However, the variable still is not isolated. The variable is being multiplied by 3. To write an equivalent equation with the variable isolated, divide by 3 on both sides.

EXAMPLE Solving Equations Using Division

Use multiplicative inverses to solve $3x = 15$.

$\dfrac{3x}{3} = \dfrac{15}{3}$ • The variable is being multiplied by 3, so use the inverse property and divide both sides by 3.

$x = 5$ • Simplify.

Check the solution.

$3(5) \overset{?}{=} 15$ • Substitute the possible solution for x.

$15 = 15$ • Because this is a true statement, 5 is the correct solution.

So, in the equation based on $3x = 15$, $x = 5$.

EXAMPLE Solving Equations Using Multiplication

Use multiplicative inverses to solve $\dfrac{n}{6} = 3$.

$\dfrac{n}{6} \times 6 = 3 \times 6$ • The variable is being divided by 6, so use the inverse property and multiply both sides by 6.

$n = 18$ • Simplify.

Check the solution.

$\dfrac{(18)}{6} \overset{?}{=} 3$ • Substitute the possible solution for n.

$3 = 3$ • Because this is a true statement, 18 is the solution.

So, in the equation $\dfrac{n}{6} = 3$, $n = 18$.

Solve each equation. Check your solution.

21 $6x = 30$

22 $\frac{y}{4} = 5$

23 $9n = -27$

24 $\frac{a}{3} = 10$

Solving Equations Requiring Two Operations

In the equation $2x - 3 = 11$, notice that the variable is being multiplied and has a term being subtracted. This type of problem can be referred to as a "two-step" equation. Still, the objective is to use equivalent equations to isolate the variable. To do this, first isolate the term that contains the variable. Then isolate the variable.

$2x - 3 = 11$	• Add 3 to both sides to isolate the term that contains the variable.
$2x - 3 + 3 = 11 + 3$	• Simplify.
$2x = 14$	• Divide by 2 on both sides to isolate the variable.
$\frac{2x}{2} = \frac{14}{2}$	• Simplify.
$x = 7$	

Check the solution.

$2x - 3 = 11$	• Substitute the possible solution for x.
$2(7) - 3 \stackrel{?}{=} 11$	• Simplify, using order of operations.
$14 - 3 \stackrel{?}{=} 11$	
$11 = 11$	• Because this is a true statement, 7 is the solution.

Solve the equation $4n + 1 = 3$.

$4n + 1 - 1 = 3 - 1$ • Subtract 1 on both sides to isolate the term containing the variable.

$4n = 2$ • Simplify.

$\dfrac{4n}{4} = \dfrac{2}{4}$ • Multiply or divide on both sides to isolate the variable.

$n = \dfrac{1}{2}$ • Check solution by substituting into original equation.

$4\left(\dfrac{1}{2}\right) + 1 \overset{?}{=} 3$ • Simplify, using order of operations.

$2 + 1 \overset{?}{=} 3$ • If the statement is true, you have the solution.

$3 = 3$

When $4n + 1 = 3$, n is $\dfrac{1}{2}$.

 Check It Out

Solve each equation. Check your solution.

25 $4x + 7 = 27$

26 $5y - 2 = 8$

27 $2n + 11 = 3$

28 $\dfrac{a}{3} + 7 = 5$

Solving Equations with the Variable on Both Sides

Consider the equation $x - 7 = -2x + 5$. Notice that both sides of the equation have a term with the variable. To solve this equation, you still have to use equivalent equations to isolate the variable.

To isolate the variable, first use the additive inverse of one of the terms that contain the variable to collect these terms on one side of the equation. (Generally, this should be on the side of the equation where the number of the variable is higher—this allows you to work with positive numbers whenever possible.) Then use the additive inverse to collect the constant terms on the other side. Then multiply or divide to isolate the variable.

EXAMPLE **Solving an Equation with Variables on Both Sides**

Solve the equation $x - 7 = -2x + 5$.

$x - 7 + 2x = -2x + 5 + 2x$	• Add $2x$ on both sides to isolate the variable.
$3x - 7 = 5$	• Simplify. Combine like terms.
$3x - 7 + 7 = 5 + 7$	• Add or subtract on both sides to collect constant terms on the side opposite the variable.
$3x = 12$	• Simplify.
$\dfrac{3x}{3} = \dfrac{12}{3}$	• Multiply or divide on both sides to isolate the variable.
$x = 4$	• Simplify.
$4 - 7 \overset{?}{=} -2(4) + 5$	• Check by substituting possible solution into original equation.
$4 - 7 \overset{?}{=} -8 + 5$	• Simplify, using order of operations.
$-3 = -3$	• If the statement is true, you substituted the correct solution.

For $x - 7 = -2x + 5$, $x = 4$.

Check It Out

Solve each equation. Check your solution.

㉙ $8n - 4 = 6n$

㉚ $12x + 4 = 15x - 2$

Equations Involving the Distributive Property

To solve the equation $3x - 4(2x + 5) = 3(x - 2) + 10$, notice that the terms are not yet ready to be collected on one side of the equation. First, you have to use the distributive property.

$3x - 8x - 20 = 3x - 6 + 10$	• Simplify, using the distributive property.
$-5x - 20 = 3x + 4$	• Combine like terms.
$-5x - 20 + 5x = 3x + 4 + 5x$	• Add or subtract on both sides to collect terms with variable on one side.
$-20 = 8x + 4$	• Combine like terms.
$-20 - 4 = 8x + 4 - 4$	• Add or subtract on both sides to collect constant terms on the side opposite from the variable.
$-24 = 8x$	• Combine like terms.
$\dfrac{-24}{8} = \dfrac{8x}{8}$	• Multiply or divide on both sides to isolate the variable.
$-3 = x$	• Simplify.
$3(-3) - 4[2(-3) + 5] \overset{?}{=} 3[(-3){-}2] + 10$ $3(-3) - 4(-6 + 5) \overset{?}{=} 3(-5) + 10$ $-9 - (-4) \overset{?}{=} -15 + 10$	• Substitute the possible solution into the original equation, and simplify using the order of operations.
$-5 = -5$	• If the statement is true, you substituted the correct solution.

Check It Out

Solve each equation. Check your solution.

31 $4(n - 2) = 12$

32 $6 - 2(x - 2) = 6(x + 3)$

APPLICATION Three Astronauts and a Cat

Here is a modern version of a problem that first appeared in the year 850.

Three astronauts and their pet cat land on a deserted asteroid that resembles Earth in many ways. They find a large lake with lots of fish in it, and they try to catch as many fish as they can before nightfall. Tired, they take shelter and decide to divide up the fish in the morning.

One astronaut wakes up during the night and decides to take her share. She divides the pile of fish into three equal parts, but there is one left over. So she gives it to the cat. She hides her third and puts the rest of the fish back in a pile. Later the second and third astronauts wake up in turn and do exactly the same thing. In the morning they divide the pile of fish that's left into three equal parts. They give the one remaining fish to the cat. What is the smallest number of fish they originally caught? See **Hot**Solutions for the answer.

5·4 Exercises

Give the additive inverse of each term.

1. 7

2. $-4x$

Determine whether the given value is the solution of the equation.

3. 6; $3(y - 2) = 12$

4. 5; $6n - 5 = 3n + 11$

Solve each equation. Check your solution.

5. $n - 6 = 11$

6. $\dfrac{y}{8} = 3$

7. $x + 12 = 7$

8. $9x = 63$

9. $\dfrac{a}{6} = -2$

10. $3x + 7 = 25$

11. $\dfrac{y}{4} - 2 = 5$

12. $2n + 11 = 7$

13. $\dfrac{a}{3} + 9 = 5$

14. $13x - 5 = 10x + 7$

15. $y + 6 = 3y - 8$

16. $8x + 6 = 3x - 4$

17. $3a + 4 = 4a - 3$

18. $6(2n - 5) = 4n + 2$

19. $9y - 4 - 6y = 2(y + 1) - 5$

20. $8x - 3(x - 1) = 4(x + 2)$

21. $14 - (6x - 5) = 5(2x - 1) - 4x$

22. Which of the following equations can be solved by adding 6 to both sides and dividing by 5 on both sides?

    A. $5x + 6 = 16$ B. $\dfrac{x}{5} + 6 = 16$

    C. $5x - 6 = 14$ D. $\dfrac{x}{5} - 6 = 14$

23. Which equation does not have $x = 4$ as its solution?

    A. $3x + 5 = 17$ B. $2(x + 2) = 10$

    C. $\dfrac{x}{2} + 5 = 7$ D. $x + 2 = 2x - 2$

5·5 Ratio and Proportion

Ratio

A **ratio** is a comparison of two quantities. If there are 12 boys and 14 girls in a class, the ratio of the number of boys to the number of girls is 12 to 14, which can be expressed as the fraction $\frac{12}{14}$ and reduced to $\frac{6}{7}$. You can write some other ratios.

Comparison	Ratio	As a Fraction
Number of girls to number of boys	14 to 12	$\frac{14}{12} = \frac{7}{6}$
Number of boys to number of students	12 to 26	$\frac{12}{26} = \frac{6}{13}$
Number of students to number of girls	26 to 14	$\frac{26}{14} = \frac{13}{7}$

Check It Out

A coin bank contains 5 nickels and 15 dimes. Write each ratio, and reduce it to the lowest terms.

1 number of nickels to number of dimes

2 number of dimes to number of coins

3 number of coins to number of nickels

Rate

A rate is a ratio that compares a quantity to one unit. Some examples of rates are listed below.

$$\frac{\$8}{1 \text{ hr}} \qquad \frac{2 \text{ cans}}{\$1} \qquad \frac{12 \text{ in.}}{1 \text{ ft}} \qquad \frac{35 \text{ mi}}{1 \text{ hr}} \qquad \frac{24 \text{ mi}}{1 \text{ gal}}$$

If a car gets $\frac{35 \text{ mi}}{1 \text{ gal}}$, then the car can get $\frac{70 \text{ mi}}{2 \text{ gal}}$, $\frac{105 \text{ mi}}{3 \text{ gal}}$, and so on. All the ratios are equal—they can be simplified to $\frac{35}{1}$.

Proportions

When two ratios are equal, they form a **proportion**. One way to determine whether two ratios form a proportion is to check their **cross products**. Every proportion has two cross products: the numerator of one ratio multiplied by the denominator of the other ratio. If the cross products are equal, the two ratios form a proportion.

EXAMPLE   Determining a Proportion

Determine whether a proportion is formed.

$$\frac{4}{5} \stackrel{?}{=} \frac{24}{40} \qquad\qquad \frac{2}{7} \stackrel{?}{=} \frac{12}{42}$$ • Find the cross products.

$$4 \times 40 \stackrel{?}{=} 5 \times 24 \qquad 2 \times 42 \stackrel{?}{=} 7 \times 12$$

$160 \neq 120$ $82 = 82$ • If the sides are equal, the

This pair is not a   This pair is a   ratios are proportional.

proportion.   proportion.

 Check It Out

Determine whether a proportion is formed.

4 $\frac{6}{9} = \frac{4}{6}$

5 $\frac{9}{36} = \frac{15}{48}$

6 $\frac{4}{12} = \frac{8}{32}$

7 $\frac{10}{25} = \frac{16}{40}$

Using Proportions to Solve Problems

To use proportions to solve problems, set up two ratios that relate what you know to what you are solving for.

Suppose that you can buy 5 DVDs for $20. How much would it cost to buy 10 DVDs? Let c represent the cost of the 10 DVDs.

If you express each ratio as $\frac{\text{DVDs}}{\$}$, then one ratio is $\frac{5}{20}$ and another is $\frac{10}{c}$. The two ratios must be equal.

$$\frac{5}{20} = \frac{10}{c}$$

To solve for c, you can use the cross products. Because you have written a proportion, the cross products are equal.

$$20 \times 10 = 5c$$
$$200 = 5c$$
$$c = 40$$

So, 10 DVDs would cost $40.00.

EXAMPLE **Solve Problems Using Proportions**

A salsa recipe calls for a tomato/onion ratio of $\frac{4}{3}$. If you used 3 cups of tomato, how many cups of onion will you need?

$\frac{4}{3} = \frac{3}{n}$ • Write the proportion.

$3 \times 3 = 4n$ • Find the cross products.

$9 = 4n$ • Divide.

$2\frac{1}{4} = n$ • Solution.

You will need $2\frac{1}{4}$ cups of onions.

 Check It Out

Use proportions to solve Exercises 8–11.

8 A car gets 30 miles per gallon. How many gallons would the car need to travel 90 miles?

9 A worker earns $100 every 8 hours. How much would the worker earn in 28 hours?

10 Carlos read 40 pages of a book in 50 minutes. How many pages should he be able to read in 80 minutes?

11 Jeannie has 8 shirts for every 5 pair of jeans. How many pairs of jeans does she have if she has 40 shirts?

5.5

RATIO AND PROPORTION

5·5 Exercises

A basketball team has 15 wins and 5 losses. Write each ratio.

1. number of wins to number of losses
2. number of wins to number of games
3. number of losses to number of games

Write each ratio, and then write it in simplest form.

4. 36 phone calls to 42 text messages
5. 50 plastic bottles to 15 recyclable cans
6. 15 hits to 45 pitches

Express each rate as a unit rate.

7. 200 miles in 5 hours
8. $350 for 5 days

Determine whether a proportion is formed. Write *yes* or *no*.

9. $\dfrac{5}{7} \overset{?}{=} \dfrac{8}{11}$

10. $\dfrac{9}{6} \overset{?}{=} \dfrac{15}{10}$

11. $\dfrac{4}{9} \overset{?}{=} \dfrac{11}{24}$

Use a proportion to solve each exercise.

12. A gallon of gasoline costs $3.89. How much would it cost to fill a 13-gallon tank?

13. At Middleton High School, there are 324 students for every 12 teachers. How many teachers are there for 405 students?

14. Your weight on Earth's moon is $\dfrac{1}{6}$ your weight on Earth. If a person weighed 120 pounds on Earth, what would be his or her weight on the moon?

5·6 Inequalities

When comparing the numbers 7 and 4, you might say "7 is greater than 4," or you might also say "4 is less than 7." When two expressions are not equal or could be equal, you can write an **inequality**. The symbols are shown in the chart.

Symbol	Meaning	Example
>	is greater than	$7 > 4$
<	is less than	$4 < 7$
≥	is greater than or equal to	$x \geq 3$
≤	is less than or equal to	$-2 \leq x$

Graphing Inequalities

The equation $x = 3$ has one solution, 3. The inequality $x > 3$ has an infinite number of solutions: 3.001, 3.2, 4, 15, 197, and 955 are just some of the solutions. Note that 3 is not a solution because 3 is not greater than 3. Because you cannot list all of the solutions, you can show them on a number line.

To show all the values that are greater than 3, but not including 3, use an open circle on 3 and shade the number line to the right.

$$x > 3$$

The inequality $y \leq -2$ also has an infinite number of solutions: -2.01, -2.5, -3, -8, and -54 are just some of the solutions. Note that -2 is also a solution because -2 is less than *or* equal to -2. On a number line, you want to show all the values that are less than or equal to -2. Because the -2 is to be included, use a closed (filled-in) circle on -2 and shade the number line to the left.

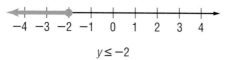

$$y \leq -2$$

 Check It Out

Draw the number line showing the solutions to each inequality.

1 $x \geq 1$ **2** $y < -3$
3 $n > -2$ **4** $x \leq 4$

Writing Inequalities

You can write equivalent inequalities just as you wrote equivalent equations. Apply what you have learned about equations to solve one-step inequalities. Inequalities are sentences that compare quantities that are not necessarily equal. Use the symbols in the chart on page 264.

EXAMPLE Writing Inequalities

Solve $s + \frac{1}{4} \leq 1\frac{1}{4}$.

$$s + \frac{1}{4} \leq 1\frac{1}{4}$$ • Write the inequality.

$$s + \frac{1}{4} - \frac{1}{4} \leq 1\frac{1}{4} - \frac{1}{4}$$ • Subtract $\frac{1}{4}$ from each side.

$$s \leq 1$$ • Simplify.

The solution is $s \leq 1$. Note that the solution is not one single number but any number that is less than or equal to 1.

Solving Inequalities

When the inequality is multiplied or divided by a negative number on each side, the inequality sign must be reversed. Begin with the inequality $3 > -1$.

$$3 > -1$$
$$-1 \times 3 \,?\, -1 \times -1$$
$$-3 < 1$$

Notice that the inequality would be false if the inequality sign had not been reversed because $-3 \not> 1$.

To determine the solutions of the inequality $-2x - 3 \geq 3$, use equivalent inequalities to isolate the variable.

$-2x - 3 \geq 3$ • Add or subtract on both sides to isolate the variable term.

$-2x - 3 + 3 \geq 3 + 3$ • Combine like terms.

$-2x \geq 6$ • Multiply or divide on both sides to isolate the variable. If you multiply or divide by a negative number, reverse the inequality sign.

$\dfrac{-2x}{-2} \leq \dfrac{6}{-2}$ • Simplify.

$x \leq -3$

Check It Out

Solve each inequality.

5 $x + 9 > 4$

6 $3n \leq -12$

7 $5y + 3 < 18$

8 $-2x + 4 \leq 2$

5·6 Exercises

Draw the number line showing the solutions to each inequality.

1. $x < -1$ 2. $y \geq 0$ 3. $n > -3$ 4. $x \leq 5$

Solve each inequality.

5. $x - 4 < 6$

6. $n + 7 > 4$

7. Which inequality has its solutions represented by the following?

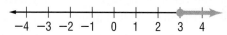

    A. $x < 3$ **B.** $x \leq 3$
    C. $x > 3$ **D.** $x \geq 3$

8. Which inequality has its solutions represented by the following?

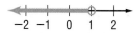

    A. $x \geq 1$ **B.** $x = 1$
    C. $x \leq 1$ **D.** $x < 1$

9. If $x = -1$, is it true that $3(x - 1) \leq 4x$?

10. If $x = 1$, is it true that $2(x - 1) < 0$?

11. Which of the following statements is false?
    A. $-7 \leq 2$ **B.** $0 \leq -4$
    C. $8 \geq -8$ **D.** $5 \geq 5$

12. Which of the following inequalities does not have $x < 2$ as its solution?
    A. $-3x < -3$ **B.** $x + 5 < 7$
    C. $4x - 1 < 7$ **D.** $-x > -2$

5·7 Graphing on the Coordinate Plane

Axes and Quadrants

When you cross a **horizontal** (left to right) number line with a **vertical** (up and down) number line, the result is a two-dimensional coordinate plane.

The number lines are called **axes**. The horizontal number line is the **x-axis**, and the vertical number line is the **y-axis**. The plane is divided into four regions, called **quadrants**. Each quadrant is named by a roman numeral, as shown in the diagram.

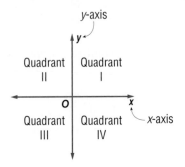

Check It Out

Fill in the blanks.

1. The horizontal number line is called the ____.
2. The lower left region of that coordinate plane is called ____.
3. The upper right region of the coordinate plane is called ____.
4. The vertical number line is called the ____.

Writing an Ordered Pair

Any location on the coordinate plane can be represented by a **point**. The location of any point is given in relation to where the two axes intersect, called the **origin**.

Two numbers are required to identify the location of a point. The x-coordinate tells you how far to the left or right of the origin the point lies. The y-coordinate tells you how far up or down from the origin the point lines. Together, the x-coordinate and y-coordinate form an **ordered pair**, (x, y).

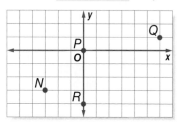

Because point Q is 6 units to the right of the origin and 1 unit up, its ordered pair is $(6, 1)$. Point N is 3 units to the left of the origin and 3 units down, so its ordered pair is $(-3, -3)$. Point R is 0 units to the left or right of the origin and 4 units down, so its ordered pair is $(0, -4)$. Point P is 0 units to the left or right of the origin and 0 units up or down. Point P is the origin, and its ordered pair is $(0, 0)$.

 Check It Out

Give the ordered pair for each point.

5 A

6 B

7 C

8 D

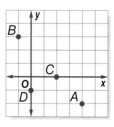

Locating Points on the Coordinate Plane

To locate point $I(2, -5)$ from the origin, move 2 units to the right and 5 units down. Point I lies in Quadrant IV. To locate point $H(-2, 4)$ from the origin, move 2 units to the left and 4 units up. Point H lies in Quadrant II. Point $J(4, 0)$ is, from the origin, 4 units to the right and 0 units up or down. Point J lies on the x-axis. Point $K(0, -1)$ is, from the origin, 0 units to the left or right and 1 unit down. Point K lies on the y-axis.

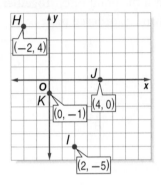

Check It Out

Draw each point on the coordinate plane and tell in which quadrant or on which axis it lies.

9 $P(2, -3)$

10 $Q(-3, 2)$

11 $R(2, 1)$

12 $S(0, 4)$

The Graph of an Equation with Two Variables

Consider the equation $y = 2x - 1$. Notice that it has two variables, x and y. Point (3, 5) is a solution of this equation. If you substitute 3 for x and 5 for y (in the ordered pair, 3 is the x-coordinate and 5 is the y-coordinate), the true statement $5 = 5$ is obtained. Point (2, 4) is not a solution of the equation. Substituting 2 for x and 4 for y results in the false statement $4 = 3$.

You can generate ordered pairs that are solutions.

Choose a value for x.	Substitute the value into the equation.	Solve for y.	Ordered Pair $y = 2x - 1$
0	$y = 2(0) - 1$	−1	(0, −1)
1	$y = 2(1) - 1$	1	(1, 1)
3	$y = 2(3) - 1$	5	(3, 5)
−1	$y = 2(-1) - 1$	−3	(−1, −3)

If you locate the points on a coordinate plane, you will notice that they all lie along a straight line.

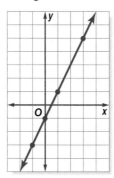

The coordinates of any point on the line will result in a true statement if they are substituted into the equation.

Graph the equation $y = \frac{1}{2}x - 1$.

• Choose five values for x.

Because the value of x is to be multiplied by $\frac{1}{2}$, choose values that are multiples of 2, such as -2, 0, 2, 4, and 6.

• Calculate the corresponding values for y.

When $x = -2$, $y = \frac{1}{2}(-2) - 1 = -2$.
When $x = 0$, $y = \frac{1}{2}(0) - 1 = -1$.
When $x = 2$, $y = \frac{1}{2}(2) - 1 = 0$.
When $x = 4$, $y = \frac{1}{2}(4) - 1 = 1$.
When $x = 6$, $y = \frac{1}{2}(6) - 1 = 2$.

• Write the five solutions as ordered pairs (x, y).

$(-2, -2)$, $(0, -1)$, $(2, 0)$, $(4, 1)$, and $(6, 2)$

• Locate the points on a coordinate plane and draw the line.

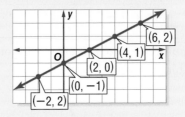

Check It Out

Find five solutions of each equation. Graph each line.

13 $y = 3x - 1$

14 $y = 2x - 1$

15 $y = \frac{1}{2}x - 2$

16 $y = -2x + 1$

Horizontal and Vertical Lines

Choose several points that lie on a horizontal line.

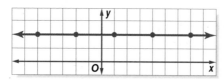

Notice that any point that lies on the line has a *y*-coordinate of 2. The equation of this line is $y = 2$.

Choose several points that lie on a vertical line.

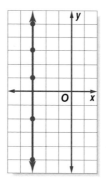

Notice that any point that lies on the line has an *x*-coordinate of −3. The equation of this line is $x = -3$.

 Check It Out

Graph each line.

17 $x = 2$

18 $y = -4$

19 $x = -3$

20 $y = 1$

5·7 Exercises

Fill in the blanks.

1. The vertical number line is called the ____.
2. The upper left region of the coordinate plane is called ____.
3. The lower right region of the coordinate plane is called ____.

Give the ordered pair for each point.

4. A
5. B
6. C
7. D

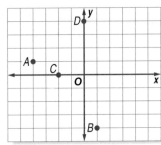

Locate each point on the coordinate plane and tell in which quadrant or on which axis it lies.

8. $H(-3, 4)$
9. $J(0, -1)$
10. $K(3, 1)$
11. $L(-2, 0)$

Find five solutions of each equation. Graph each line.

12. $y = 2x - 1$
13. $y = -3x + 2$
14. $y = \frac{1}{2}x - 2$

5·8 Slope and Intercept

Slope

One characteristic of a line is its **slope**. Slope is a measure of a line's steepness. To describe the way a line slants, you need to observe how the coordinates on the line change as you move right. The relationship of one quantity to another is called the **rate of change**. A rate of change is usually expressed as a unit rate. Choose two points along the line. The run is the difference in the x-coordinates. The rise is the difference in the y-coordinates.

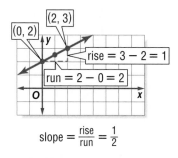

$$\text{slope} = \frac{\text{rise}}{\text{run}} = \frac{1}{2}$$

The slope, then, is given by the ratio of rise (vertical movement) to run (horizontal movement).

$$\text{Slope} = \frac{\text{rise}}{\text{run}}$$

Notice that for line a, the rise between the two marked points is 10 units and the run is 4 units. The slope of the line, then, is $\frac{10}{4} = \frac{5}{2}$. For line b, the rise is -3 and the run is 6, so the slope of the line is $-\frac{3}{6} = -\frac{1}{2}$.

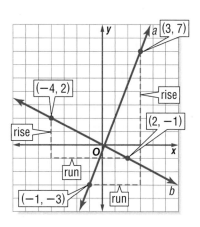

The slope along a straight line is always the same. For line a, regardless of the two points chosen, the slope will always simplify to $\frac{5}{2}$.

Determine the slope of each line.

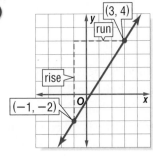

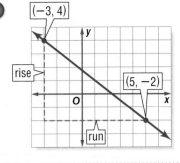

Calculating the Slope of a Line

You can calculate the slope of a line just from knowing two points on the line. The rise is the difference of the *y*-coordinates and the run is the difference of the *x*-coordinates. For the line that passes through the points (1, −2) and (4, 5), the slope can be calculated as shown. The variable *m* is used to represent slope.

$$m = \frac{\text{rise}}{\text{run}} = \frac{5 - (-2)}{4 - 1} = \frac{7}{3}$$

The slope could also have been calculated another way.

$$m = \frac{\text{rise}}{\text{run}} = \frac{-2 - 5}{1 - 4} = \frac{-7}{-3} = \frac{7}{3}$$

The order of the coordinates does not matter, as long as the order is consistent between the rise and the run.

Find the slope of the line that contains the points
(−2, 3) and (4, −1).

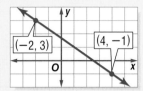

$m = \dfrac{-1 - 3}{4 - (-2)}$ or $m = \dfrac{3 - (-1)}{-2 - 4}$ • Use the definition

$$m = \frac{\text{rise}}{\text{run}} = \frac{\text{difference of } y\text{-coordinates}}{\text{difference of } x\text{-coordinates}}$$
to find the slope.

$m = \dfrac{-4}{6}$ or $m = \dfrac{4}{-6}$ • Simplify.

$m = \dfrac{-2}{3}$ or $m = \dfrac{2}{-3}$

The slope is $-\dfrac{2}{3}$.

Check It Out

Find the slope of the line that contains the given points.

3 (−2, 7) and (4, 1)

4 (−1, −2) and (3, 4)

5 (0, 3) and (6, 0)

6 (−1, −3) and (1, 5)

Slopes of Horizontal and Vertical Lines

Choose two points on a horizontal line, $(-1, 2)$ and $(3, 2)$. Calculate the slope of the line.

$$m = \frac{\text{rise}}{\text{run}} = \frac{2 - 2}{3 - (-1)} = \frac{0}{4} = 0$$

A horizontal line has no rise; its slope is 0.

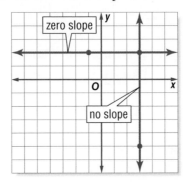

Choose two points on a vertical line, $(3, 2)$ and $(3, -5)$. Calculate the slope of the line.

$$m = \frac{\text{rise}}{\text{run}} = \frac{-5 - 2}{3 - 3} = \frac{-7}{0}, \text{ which is undefined.}$$

A vertical line has no run; it has *no slope*.

 Check It Out

Find the slope of the line that contains the given points.

7 $(-2, 3)$ and $(4, 3)$ **8** $(1, -2)$ and $(1, 5)$

9 $(-4, 0)$ and $(-4, 6)$ **10** $(4, -2)$ and $(-1, -2)$

The *y*-Intercept

A second characteristic of a line, after the slope, is the **y-intercept**. The *y*-intercept is the location along the *y*-axis where the line crosses, or intercepts, the *y*-axis.

The *y*-intercept of line *a* is 2, and the *y*-intercept of line *b* is −4.

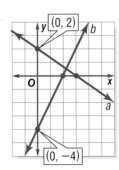

Identify the *y*-intercept of each line.

 11 *c*

12 *d*

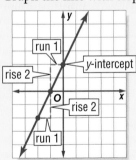

Using the Slope and *y*-Intercept to Graph a Line

A line can be graphed if the slope and the *y*-intercept are known. First you locate the *y*-intercept on the *y*-axis. Then you use the rise and the run of the slope to locate a second point on the line. Connect the two points to plot your line.

EXAMPLE **Using the Slope and *y*-Intercept to Graph a Line**

Graph the line with slope 2 and *y*-intercept 2.

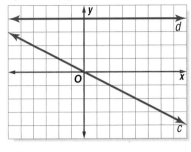

- Locate the *y*-intercept.
- Use the slope to locate other points on the line. If the slope is a whole number *m*, remember $m = \frac{m}{1}$, so rise is *m* and run is 1.
- Draw a line through the points.

Check It Out

Graph each line.

⑬ slope $= \frac{1}{3}$; y-intercept at -1

⑭ slope $= -\frac{2}{5}$; y-intercept at 2

⑮ slope $= 2$; y-intercept at -1

⑯ slope $= -3$; y-intercept at 0

Slope-Intercept Form

The equation $y = mx + b$ is in the slope-intercept form for the equation of a line. When an equation is in this form, the slope of the line is given by m and the y-intercept is located at b. The graph of the equation $y = \frac{2}{3}x - 4$ is a line that has a slope of $\frac{2}{3}$ and a y-intercept at -4. The graph is shown.

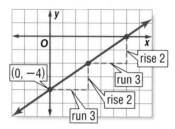

 Check It Out

Determine the slope and the y-intercept from the equation of each line.

⑰ $y = -3x + 2$

⑱ $y = \frac{1}{4}x - 2$

⑲ $y = -\frac{2}{3}x$

⑳ $y = 6x - 5$

5.8

SLOPE AND INTERCEPT

5·8 Exercises

Determine the slope of each line for Exercises 1–6.

1.

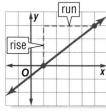

2.

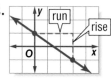

3. contains $(-1, 3)$ and $(-1, 0)$

4. contains $(-4, 2)$ and $(4, -2)$

5. contains $(0, -3)$ and $(2, 6)$

6. contains $(5, -2)$ and $(5, 4)$

Graph each line.

7. slope $= -\frac{1}{3}$; y-intercept at 3

8. slope $= 3$; y-intercept at -4

9. slope $= -1$; y-intercept at 2

10. slope $= 0$; y-intercept at 1

Determine the slope and the y-intercept from the equation of each line.

11. $y = -2x - 3$

12. $y = \frac{3}{4}x + 2$

13. $y = x + 1$

14. $y = -4$

15. $x = 3$

Algebra

You can use the problems and the list of words that follow to see what you learned in this chapter. You can find out more about a particular problem or word by referring to the topic number (*for example,* Lesson 5·2).

Problem Set

Write an equation for each sentence. (Lesson 5·1)

1. If 5 is subtracted from the product of 3 and a number, the result is 7 more than the number.

2. 4 times the sum of a number and 3 is 6 less than twice the number.

Factor out the greatest common factor in each expression. (Lesson 5·2)

3. $5x + 35$

4. $8n - 6$

Simplify each expression. (Lesson 5·2)

5. $8a - 3b - a + 7b$

6. $4(2n - 1) - (2n + 3)$

7. Find the distance traveled by an in-line skater who skates at 10 miles per hour for $1\frac{1}{2}$ hours. Use the formula $d = rt$. (Lesson 5·3)

Solve each equation. Check your solution. (Lesson 5·4)

8. $x + 7 = 12$

9. $\frac{y}{3} = -5$

10. $6n - 7 = 2n + 9$

11. $y - 3 = 7y + 9$

12. $7(n - 2) = 2n + 6$

13. $10x - 3(x - 1) = 4(x + 3)$

Use a proportion to solve.

14. A map is drawn using a scale of 60 miles to 1 centimeter. On the map, the two cities are 5.5 centimeters apart. What is the actual distance between the two cities? (Lesson 5·5)

Solve each inequality. Graph the solution. (Lesson 5·6)

15. $x + 7 \leq 5$ **16.** $2x + 6 > 2$

Locate each point on the coordinate plane and tell where it lies.

17. $A(3, 1)$ **18.** $B(-2, 0)$ **19.** $C(-1, -4)$ **20.** $D(0, 3)$

21. Find the slope of the line that contains the points $(5, -2)$ and $(-1, 2)$.

Determine the slope and the y-intercept from the equation of each line. Graph the line.

22. $y = \frac{1}{2}x + 2$ **23.** $y = -1$

24. $x + 2y = 4$ **25.** $4x + 2y = 0$

HotWords

Write definitions for the following words.

additive inverse (Lesson 5·4)

associative property (Lesson 5·2)

axes (Lesson 5·7)

Celsius (Lesson 5·3)

commutative property (Lesson 5·2)

cross product (Lesson 5·5)

difference (Lesson 5·1)

distributive property (Lesson 5·2)

equation (Lesson 5·1)

equivalent (Lesson 5·1)

equivalent expression (Lesson 5·2)

expression (Lesson 5·1)

Fahrenheit (Lesson 5·3)

formula (Lesson 5·3)

horizontal (Lesson 5·7)

inequality (Lesson 5·6)

like terms (Lesson 5·2)

order of operations (Lesson 5·3)

ordered pair (Lesson 5·7)

origin (Lesson 5·7)

perimeter (Lesson 5·3)

point (Lesson 5·7)

product (Lesson 5·1)

proportion (Lesson 5·5)

quadrant (Lesson 5·7)

quotient (Lesson 5·1)

rate (Lesson 5·5)

rate of change (Lesson 5·8)

ratio (Lesson 5·5)

slope (Lesson 5·8)

solution (Lesson 5·4)

sum (Lesson 5·1)

term (Lesson 5·1)

variable (Lesson 5·1)

vertical (Lesson 5·7)

x-axis (Lesson 5·7)

y-axis (Lesson 5·7)

y-intercept (Lesson 5·8)

HotTopic 6

Geometry

What do you know?

You can use the problems and the list of words that follow to see what you already know about this chapter. The answers to the problems are in **HotSolutions** at the back of the book, and the definitions of the words are in **HotWords** at the front of the book. You can find out more about a particular problem or word by referring to the topic number (*for example,* Lesson 6·2).

Problem Set

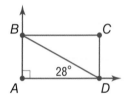

1. Find $m\angle ABD$. (Lesson 6·1)

2. If $ABCD$ is a rectangle, what is $m\angle DBC$? (Lesson 6·2)

3. Which of the letters below have more than one line of symmetry? H E X A G O N (Lesson 6·3)

4. What is the length of each side of a square that has a perimeter of 56 centimeters? (Lesson 6·4)

5. What is the area of a triangle with a base of 15 feet and a height of 8 feet? (Lesson 6·5)

6. The bases of a trapezoid measure 6 inches and 10 inches. Its height is 7 inches. Find the area of the trapezoid. (Lesson 6·5)

7. A cylinder with a radius of 10 meters is 10 meters high. Find the surface area of the cylinder. Use $\pi \approx 3.14$. (Lesson 6·6)

8. Find the volume of a cube whose sides measure 5 inches. (Lesson 6·7)

9. The triangular faces of a triangular prism each have an area of 10 square centimeters. The prism is 8 centimeters long. What is its volume? (Lesson 6·7)

10. What is the area of circle *B* in terms of π?

(Lesson 6·8)

11. Find the length of $\overline{QS}$. (Lesson 6·9)

12. Find the length of $\overline{QR}$. (Lesson 6·9)

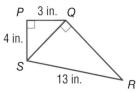

HotWords

acute angle (Lesson 6·1)

angle (Lesson 6·1)

circumference (Lesson 6·6)

complementary angle
  (Lesson 6·1)

congruent angle (Lesson 6·1)

cube (Lesson 6·2)

cylinder (Lesson 6·6)

degree (Lesson 6·1)

diagonal (Lesson 6·2)

diameter (Lesson 6·8)

face (Lesson 6·2)

hypotenuse (Lesson 6·9)

isosceles triangle (Lesson 6·1)

legs of a triangle (Lesson 6·5)

line of symmetry (Lesson 6·3)

opposite angle (Lesson 6·2)

parallelogram (Lesson 6·2)

perimeter (Lesson 6·4)

perpendicular (Lesson 6·5)

pi (Lesson 6·8)

point (Lesson 6·1)

polygon (Lesson 6·1)

polyhedron (Lesson 6·2)

prism (Lesson 6·2)

Pythagorean Theorem
  (Lesson 6·4)

Pythagorean triple (Lesson 6·9)

quadrilateral (Lesson 6·2)

radius (Lesson 6·8)

ray (Lesson 6·1)

rectangular prism (Lesson 6·2)

reflection (Lesson 6·3)

regular polygon (Lesson 6·2)

regular shape (Lesson 6·2)

right angle (Lesson 6·1)

right triangle (Lesson 6·4)

rotation (Lesson 6·3)

segment (Lesson 6·8)

supplementary angle
  (Lesson 6·1)

surface area (Lesson 6·6)

tetrahedron (Lesson 6·2)

transformation (Lesson 6·3)

translation (Lesson 6·3)

trapezoid (Lesson 6·2)

triangular prism (Lesson 6·6)

vertex (Lesson 6·1)

vertical angle (Lesson 6·1)

volume (Lesson 6·7)

6·1 Naming and Classifying Angles and Triangles

Points, Lines, and Rays

In the world of math, it is sometimes necessary to refer to a specific **point** in space. Simply draw a small dot with a pencil to represent a point. A point has no size; its only function is to show position.

To name a point, use a single capital letter.

• M

Point *M*

If you draw two points on a sheet of paper, a *line* can be used to connect them. Imagine this line as being perfectly straight and continuing without end in opposite directions. It has no thickness.

To name a line, use any two points on the line.

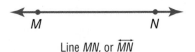

Line *MN*, or $\overleftrightarrow{MN}$

A **ray** is part of a line that extends without end in one direction. In $\overrightarrow{MN}$, which is read as "ray *MN*," *M* is the endpoint. The second point that is used to name the ray can be any point other than the endpoint. You could also name this ray $\overrightarrow{MO}$.

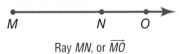

Ray *MN*, or $\overrightarrow{MO}$

Look at the line below.

K   L

1 Name the line in two different ways.

2 What is the endpoint of $\overrightarrow{KL}$?

Naming Angles

Imagine two different rays with the same endpoint. Together they form what is called an **angle**. The point they have in common is called the **vertex** of the angle. The rays form the sides of the angle.

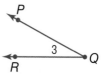

The angle above is made up of $\overrightarrow{QP}$ and $\overrightarrow{QR}$. Q is the common endpoint of the two rays. Point Q is the vertex of the angle. Instead of writing the word *angle*, you can use the symbol for an angle, which is ∠.

There are several ways to name an angle. You can name it using the three letters of the points that make up the two rays with the vertex as the middle letter (∠PQR, or ∠RQP). You can also use just the letter of the vertex to name the angle (∠Q). Sometimes you might want to name an angle with a number (∠3).

When more than one angle is formed at a vertex, you use three letters to name each of the angles. Because G is the vertex of three different angles, each angle needs three letters to name it: ∠DGF; ∠DGE; ∠EGF.

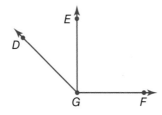

Check It Out

Look at the angles formed by the
rays at the right.

3 Name the vertex.

4 Name all the angles.

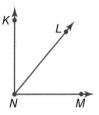

Measuring Angles

You measure an angle in **degrees**, using a *protractor* (p. 379).
The number of degrees in an angle will be greater than 0 and
less than or equal to 180.

EXAMPLE Measuring with a Protractor

Measure $\angle ABC$.

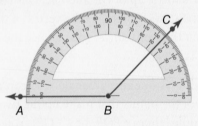

- Place the center point of the protractor on the vertex of the angle. Align the
  0° line on the protractor with one side of the angle.
- Read the number of degrees on the scale where it intersects the second side
  of the angle.

So, $m\angle ABC = 135°$.

Check It Out

Measure the angles,
using a protractor.

5 $\angle PSQ$

6 $\angle QSR$

7 $\angle PSR$

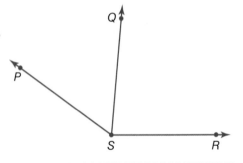

Classifying Angles

You can classify angles by their measures.

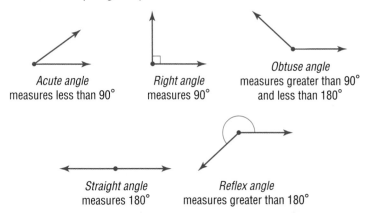

Acute angle
measures less than 90°

Right angle
measures 90°

Obtuse angle
measures greater than 90°
and less than 180°

Straight angle
measures 180°

Reflex angle
measures greater than 180°

Angles that share a side are called *adjacent angles*. You can add measures if the angles are adjacent.

$m\angle KNL = 25°$

$m\angle LNM = 65°$

$m\angle KNM = 25° + 65° = 90°$

Because the sum is 90°, you know that $\angle KNM$ is a **right angle**.

Check It Out

Use a protractor to measure and classify each angle.

8 $\angle SQR$

9 $\angle PQR$

10 $\angle PQS$

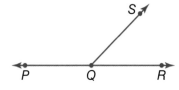

Special Pairs of Angles

When the sum of two angles equals 180°, they are called
supplementary angles.

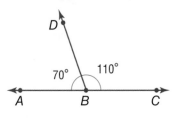

Supplementary

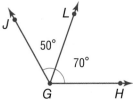

Not Supplementary

$\angle ABD = 70°$

$\angle DBC = 110°$

$70° + 110° = 180°$

$\angle ABD + \angle DBC = 180°$

$\angle JGL = 50°$

$\angle LGH = 70°$

$50° + 70° = 120°$

$\angle JGL + \angle LGH = 120°$

Opposite angles formed by two intersecting lines are called
vertical angles.

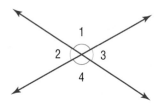

$\angle 1$ and $\angle 4$ are vertical angles.

$\angle 2$ and $\angle 3$ are vertical angles.

Check It Out

Identify each pair of angles as supplementary or vertical.

11

12 $\angle 6$ and $\angle 8$

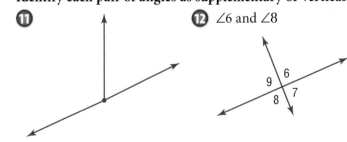

When two angles have the same angle measure, they are called **congruent angles**. ∠*FTS* and ∠*GPQ* are congruent because each angle measures 45°.

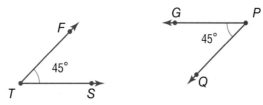

If the sum of the measure of two angles is 90°, then the angles are **complementary angles**.

Complementary   Not Complementary

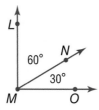   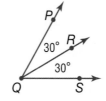

∠*NMO* = 30°   ∠*PQR* = 30°
∠*LMN* = 60°   ∠*RQS* = 30°
30° + 60° = 90°   30° + 30° = 60°
∠*NMO* + ∠*LMN* = 90°   ∠*PQR* + ∠*RQS* = 60°

Check It Out

Identify each pair of angles as complementary or congruent.

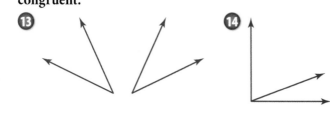

⓭   ⓮

Triangles

Triangles are **polygons** (p. 298) that have three sides, three vertices, and three angles.

You name a triangle using the three vertices in any order. $\triangle ABC$ is read "triangle ABC."

Classifying Triangles

Like angles, triangles are classified by their angle measures. They are also classified by the number of *congruent* sides, which are sides with equal length.

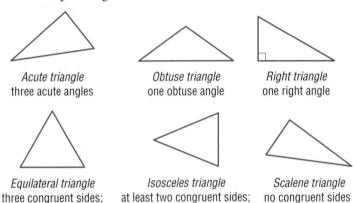

Acute triangle
three acute angles

Obtuse triangle
one obtuse angle

Right triangle
one right angle

Equilateral triangle
three congruent sides;
three congruent angles

Isosceles triangle
at least two congruent sides;
at least two congruent angles

Scalene triangle
no congruent sides

The sum of the measures of the three angles in a triangle is always 180°.

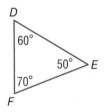

In $\triangle DEF$, $m\angle D = 60°$, $m\angle E = 50°$, and $m\angle F = 70°$.

$60° + 50° + 70° = 180°$

So, the sum of the angles of $\triangle DEF$ is 180°.

EXAMPLE | Finding the Measure of the Unknown Angle in a Triangle

∠S is a right angle, so its measure is 90°. The measure of ∠T is 35°. Find the measure of ∠U.

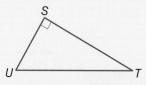

$90° + 35° = 125°$	• Add the two known angles.
$180° - 125° = 55°$	• Subtract the sum from 180°.
So, $\angle U = 55°$.	• The difference is the measure of the third angle.

Check It Out

Find the measure of the third angle of each triangle.

 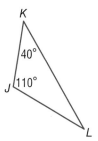

15 K

40°

J 110°

L

 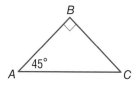

16 B

A 45°

C

17 D

60°

E 60° F

6·1 Exercises

Use the figure to answer Exercises 1–5.

1. Give six names for the line that passes through point *P*.

2. Name four rays that begin at point *Q*.

3. Name the right angle.

4. Find $m\angle PQT$.

5. Find $m\angle PQS$.

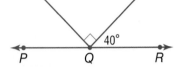

Use the figure below to answer Exercises 6–8.

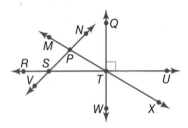

6. Identify a pair of complementary angles.

7. Identify a pair of supplementary angles.

8. Identify a pair of vertical angles.

Use the figure to answer Exercises 9 and 10.

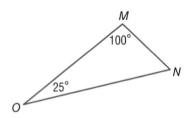

9. Is △*MNO* an acute, an obtuse, or a right triangle?

10. Is △*MNO* a scalene, an isosceles, or an equilateral triangle?

6·2 Polygons and Polyhedrons

Quadrilaterals

Quadrilaterals are polygons that have four sides and four angles. The sum of the angles of a quadrilateral is 360°. Some special quadrilaterals are classified by their sides and angles.

To name a quadrilateral, list the four vertices, either clockwise or counterclockwise.

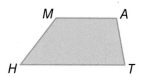

Angles of a Quadrilateral

The sum of the angles of a quadrilateral is 360°. If you know the measures of three angles in a quadrilateral, you can find the measure of the fourth angle.

EXAMPLE	Finding the Measure of the Unknown Angle in a Quadrilateral

Find $m\angle P$ in quadrilateral PQRS.

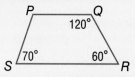

$120° + 60° + 70° = 250°$ • Add the measures of the three known angles.

$360° - 250° = 110°$ • Subtract the sum from 360°.

The difference is the measure of the fourth angle.

So, $m\angle P = 110°$.

Check It Out

Use the figure to answer Exercises 1–3.

1 Name the quadrilateral in two ways.

2 What is the sum of the angles of a quadrilateral?

3 Find $m\angle C$.

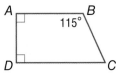

Types of Quadrilaterals

A rectangle is a quadrilateral with four right angles. *EFGH* is a rectangle. Its length is 6 centimeters, and its width is 4 centimeters.

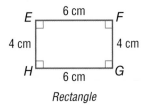

Rectangle

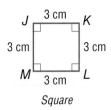

Square

Opposite sides of a rectangle are congruent. If all four sides of the rectangle are congruent, the rectangle is called a *square*. A square is a **regular polygon** because all of its sides are congruent and all of the interior angles are congruent. Some rectangles may be squares, but *all* squares are rectangles. So *JKLM* is both a square and a rectangle.

A **parallelogram** is a quadrilateral with opposite sides that are *parallel*. In a parallelogram, opposite sides are congruent, and **opposite angles** are congruent. Both *NOPQ* and *RSTU* are parallelograms.

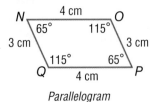

Parallelogram

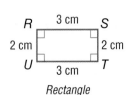

Rectangle

POLYGONS AND POLYHEDRONS

6·2

Some parallelograms may be rectangles, but *all* rectangles are parallelograms. Therefore squares are also parallelograms. If all four sides of a parallelogram are the same length, the parallelogram is called a *rhombus*. Both *WXYZ* and *JKLM* are rhombi.

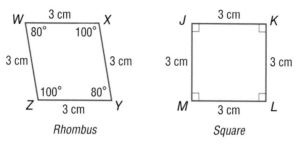

Rhombus   Square

Every square is a rhombus, although not every rhombus is a square, because a square also has equal angles.

In a **trapezoid**, two sides are parallel and two are not. A trapezoid is a quadrilateral, but it is not a parallelogram. *SNAP* is a trapezoid.

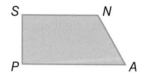

 Check It Out

Use the figure to answer Exercise 4.

④ Is quadrilateral *STUV* a rectangle? a parallelogram? a square? a rhombus? a trapezoid?

⑤ Is a square a rhombus? Why or why not?

Polygons

A polygon is a closed figure that has three or more sides. Each side is a line segment, and the sides meet only at the endpoints, or vertices.

This figure is a polygon.   These figures are not polygons.

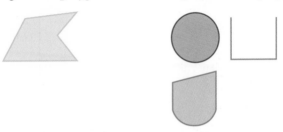

A rectangle, a square, a parallelogram, a rhombus, a trapezoid, and a triangle are all examples of polygons.

There are some aspects of polygons that are always true. For example, a polygon of *n* sides has *n* angles and *n* vertices. A polygon with three sides has three angles and three vertices. A polygon with eight sides has eight angles and eight vertices, and so on.

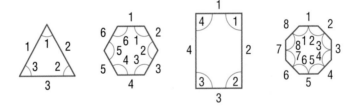

A line segment connecting two vertices of a polygon is either a side or a **diagonal**. $\overline{UT}$ is a side of polygon *PQRSTU*. $\overline{PT}$ is a diagonal.

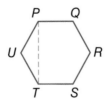

A polygon is classified by its number of sides.

Types of Polygons		
Name	**Number of Sides**	**Model**
Triangle	3	
Quadrilateral	4	
Pentagon	5	
Hexagon	6	
Heptagon	7	
Octagon	8	
Nonagon	9	
Decagon	10	

A **regular polygon** has all sides congruent and all angles congruent. Equilateral triangles and squares are examples of regular polygons.

State whether the figure is a polygon. If it is a polygon, classify it according to the number of sides it has.

6    **7**   **8**

POLYGONS AND POLYHEDRONS

6·2

APPLICATION **A Tangram Zoo**

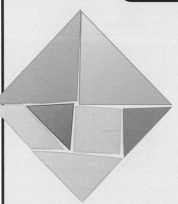

A *tangram* is an ancient Chinese puzzle. The seven tangram pieces fit together to form a square. To make your own set of tangram pieces, you can cut the shapes, as shown, from a piece of heavy paper or cardboard. You will need a ruler and a protractor, so you can draw right angles and measure the sides as necessary. The areas of the two large triangles are each one-quarter the area of the large square. The area of the small square is one-eighth the area of the large square. And the two small triangles each have one-half the area of the small square.

See if you can use all seven tangram pieces to make this tangram animal. Then see what other tangram pictures you can make. See **HotSolutions** for the answer.

Angles of a Polygon

The sum of the angles of *any* polygon totals at least 180° (triangle). Each additional side adds 180° to the measure of the first three angles. To see why, look at a *pentagon*.

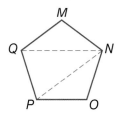

Drawing diagonals $\overline{QN}$ and $\overline{PN}$ shows that the sum of the angles of a pentagon is the sum of the angles in three triangles.

$$3 \times 180° = 540°$$

So the sum of the angles of a pentagon is 540°.

You can use the formula $(n - 2) \times 180°$ to find the sum of the angles of an *n*-sided polygon.

EXAMPLE **Finding the Sum of the Angles of a Hexagon**

Find the sum of the angles of a hexagon.

$(n - 2) \times 180 =$	• Use the formula.
$(6 - 2) \times 180 =$	• Substitute the number of sides for *n*.
$4 \times 180 = 720°$	• Use order of operations.

So, the sum of the angles of a hexagon is 720°.

To find the measure of each angle of a regular polygon, divide the sum of the angles by the number of angles in the polygon.

EXAMPLE Finding the Measure of the Angles of a Polygon

Find the measure of each angle in a regular octagon.

$(n - 2) \times 180 =$ • Use the formula.

$(8 - 2) \times 180 = 1,080$ • Substitute the number of sides for n and solve.

$1,080 \div 8 = 135$ • Divide by the number of angles.

So, each angle of a regular octagon measures 135°.

Check It Out

Solve.

9 Find the sum of the angles of a polygon with 7 sides.

10 Find the measure of each angle in a regular hexagon.

Polyhedrons

Some solid shapes are curved, like these. These shapes are not polyhedrons.

Sphere

Cylinder

Cone

Other solid shapes have flat surfaces. Each of the figures below is a **polyhedron**.

Cube

Prism

Pyramid

A polyhedron is any solid whose surface is made of polygons. The polygons are the **faces** of the polyhedrons.

A **prism** has two bases, or "end" faces. The bases of a prism are the same size and shape and are parallel to each other. The other faces are parallelograms. The bases of each prism below are shaded. When all six faces of a **rectangular prism** are square, the figure is a **cube**.

Prisms

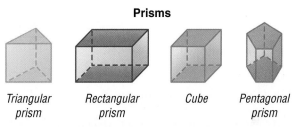

| Triangular prism | Rectangular prism | Cube | Pentagonal prism |

A *pyramid* is a structure that has one base in the shape of a polygon. It has triangular faces that meet a point called the *apex*. The base of each pyramid below is shaded. A triangular pyramid is a **tetrahedron**. A tetrahedron has four faces. Each face is triangular.

Pyramids

| Triangular pyramid | Rectangular pyramid | Square pyramid | Pentagonal pyramid |

Check It Out

Identify each polyhedron.

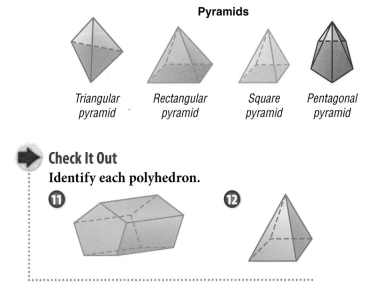

11.

12.

6·2 Exercises

Use the figures to answer the exercises.

1. Name the quadrilateral in two different ways.

2. Find $m\angle S$.

3. Is *PQRS* a parallelogram? Explain.

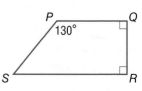

ABCD **is a parallelogram.**

4. What is the length of $\overline{BC}$?

5. Find $m\angle B$.

6. Is *ABCD* a rectangle? Explain.

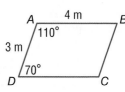

Tell whether each statement below is *true* or *false*.

7. A rectangle cannot be a rhombus.

8. Every rhombus is a parallelogram.

9. Every square is a rhombus, a rectangle, and a parallelogram.

10. Every parallelogram is a rectangle.

11. Every rectangle is a parallelogram.

12. Every trapezoid is a quadrilateral.

13. Every quadrilateral is a trapezoid.

Identify each polygon.

14.

15.

16.

17.

18.

19.

20. What is the sum of the angles in a polygon with 9 sides?

21. What is the measure of each angle in a regular pentagon?

Identify each polyhedron.

22.

23.

24.

25.

26.

27.

Refer to the figure below for Exercises 28–30.

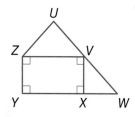

28. Name a trapezoid in the figure.

29. Name a rectangle in the figure.

30. Name a pentagon in the figure.

6·3 Symmetry and Transformations

Whenever you move a shape that is in a plane, you are performing a **transformation**. There are three basic types of transformation: reflections, rotations, and translations.

Reflections

A **reflection** (or **flip**) is one kind of transformation. When you hear the word "reflection," you may think of a mirror. The mirror image, or reverse image, of a point, a line, or a shape is called a *reflection*.

The reflection of a point is another point on the other side of a **line of symmetry**. Both the point and its reflection are the same distance from the line. Also, the segment that connects the points is perpendicular to the line of symmetry.

T' reflects point T on the other side of line *l*. T' is read "*T*-prime." T' is called the *image* of T.

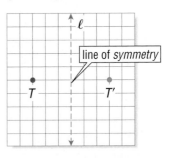

Quadrilateral *WXYZ* is reflected on the other side of line *m*. The image of *WXYZ* is *W'X'Y'Z'*.

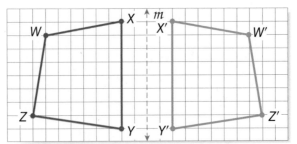

To find the reflection of a shape, measure the horizontal distance of each point to the line of symmetry. The image of each point will be the same horizontal distance from the line of symmetry on the opposite side.

In the quadrilateral reflection on the previous page, point W is 8 units from the line of symmetry, and point W' is also 8 units from the line on the opposite side. You can measure the distance from the line to each point. The corresponding image point will be the same distance from the line.

Check It Out

Look at the figure below.

1 Copy the figure on grid paper. Then draw and label the reflection.

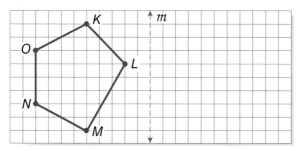

APPLICATION Flip It, Slide It, Turn It Messages

Can you read this secret message?
See **HotSolutions** for the answer.

CAN MATH BE FUN?

Use flips, slides, or turns with the letters of the alphabet and with numbers to send a secret message to a friend. Make up problems for a classmate or partner to solve, using different operations, but write them in "code." See if your partner can solve them.

Reflection Symmetry

A line of symmetry can also *separate* a shape into two parts, where one part is a reflection of the other. Each of these figures is symmetrical with respect to the line of symmetry.

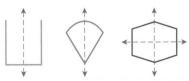

Sometimes a figure has more than one line of symmetry. A rectangle has two lines of symmetry. A square has four lines of symmetry. A circle has an infinite number of lines of symmetry because any line through the center of a circle is a line of symmetry.

A rectangle has two lines of *symmetry.*

A square has four lines of *symmetry.*

Any line through the center of a circle is a line of *symmetry.* So a circle has an infinite number of lines of symmetry.

 Check It Out

Tell whether each figure has reflection symmetry. If your answer is *yes,* tell how many lines of symmetry can be drawn through the figure.

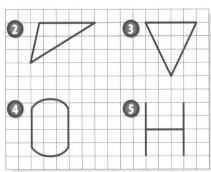

Rotations

A **rotation** (or **turn**) is a transformation that turns a line or a shape around a fixed point called the *center of rotation*. The number of degrees of rotation of a shape is usually measured in the counterclockwise direction.

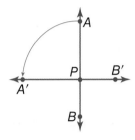

$\overleftrightarrow{AB}$ is rotated 90° around point *P*.

If a figure is rotated 360°, its position is unchanged. In the figure below, $\overrightarrow{ST}$ is rotated 360° around point *M*. $\overrightarrow{ST}$ is still in the same place.

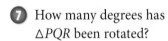

 Check It Out

Find the degrees of rotation.

6 How many degrees has $\overrightarrow{DE}$ been rotated?

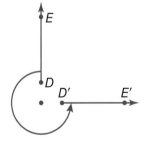

7 How many degrees has △*PQR* been rotated?

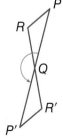

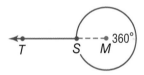

Translations

A **translation** (or **slide**) is another kind of transformation. When you move a figure to a new position without turning it, you are performing a translation.

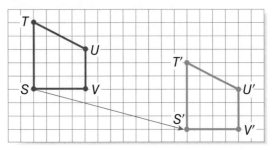

Trapezoid $S'T'U'V'$ is the image of $STUV$ under a translation. S' is 12 units to the right and 3 units down from S. All other points on the trapezoid have moved the same way.

 Check It Out

Does each pair of figures represent a translation? If you write *yes*, describe the translation.

8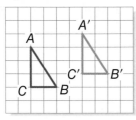

9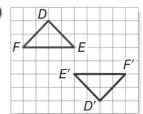

10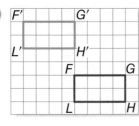

6•3 Exercises

Copy the shape below on grid paper. Then draw and label the reflection across line *a*.

1.

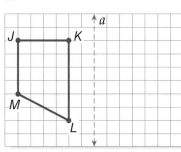

2.

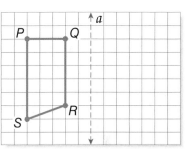

Copy the shapes. Then draw all the lines of symmetry for each.

3.   4.   5.

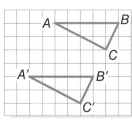

Find the number of degrees that each figure has been rotated about point *P*.

6.

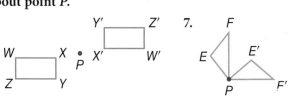

7.

Find the direction and number of units that each figure has been moved in the translations.

8.

9.

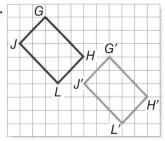

6·4 Perimeter

Perimeter of a Polygon

Dolores is planning to put a frame around a painting. To determine how much framing she needs, she must calculate the **perimeter** of, or *distance around*, the painting.

The perimeter of any polygon is the sum of the lengths of the sides of the polygon.

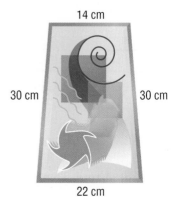

14 cm

30 cm   30 cm

22 cm

$$P = 30 \text{ cm} + 30 \text{ cm} + 14 \text{ cm} + 22 \text{ cm} = 96 \text{ cm}$$

The perimeter of the painting is 96 centimeters. Dolores will need 96 centimeters of framing.

EXAMPLE   **Finding the Perimeter of a Polygon**

Find the perimeter.

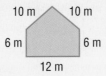

10 m   10 m

6 m   6 m

12 m

- Add the lengths of all the sides.

$$P = 6 \text{ m} + 10 \text{ m} + 10 \text{ m} + 6 \text{ m} + 12 \text{ m} = 44 \text{ m}$$

So, the perimeter of this pentagon is 44 m.

Regular Polygon Side Lengths

You can find the side lengths of a regular polygon if you know the perimeter and the total number of sides of the figure. To find the length of a side, divide the perimeter by the total number of sides. The quotient is equal to the length of one side.

PERIMETER

6·4

If a regular hexagon has a perimeter of 15 centimeters, divide the perimeter by 6.

$$15 \div 6 = 2.5$$

The length of each side is 2.5 centimeters.

 Check It Out

Find the perimeter of each polygon.

1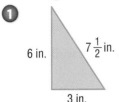

2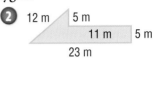

Find the length of each side.

3 a regular octagon with a perimeter of 32 feet

4 an equilateral triangle with a perimeter of 63 meters

Perimeter of a Rectangle

To find the perimeter of a rectangle, you need to know only its length and width because the opposite sides are equal.

For a rectangle with length, ℓ, and width, w, the perimeter, P, can be found with the formula $P = 2\ell + 2w$.

EXAMPLE Finding the Perimeter of a Rectangle

Find the perimeter of a rectangle with a length of 11 centimeters and a width of 8 centimeters.

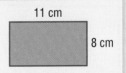

$P = 2\ell + 2w$ • Use the perimeter formula.

$= (2 \times 11) + (2 \times 8)$ • Substitute the values for ℓ and w.

$= 22 + 16 = 38$ cm • Use order of operations to solve.

So, the perimeter is 38 centimeters.

A square is a rectangle whose length and width are equal. So the formula for finding the perimeter of a square, whose sides measure s, is $P = 4 \times s$ or $P = 4s$.

Check It Out

Find the perimeter of each polygon.

5 rectangle with length 10 feet and width 5 feet

6 square with sides of 9 inches

7 rectangle with length 6 centimeters and width 2 centimeters

---

APPLICATION **The Pentagon**

Located near Washington, D.C., the Pentagon is one of the largest office buildings in the world. The United States Army, Navy, and Air Force all have their headquarters there.

The building covers an area of 29 acres and has 3,707,745 ft² of usable office space.

The structure consists of five concentric regular pentagons with ten spokelike corridors connecting them. The outside perimeter of the building is about 4,620 ft. What is the length of an outermost side? See **HotSolutions** for the answer.

PERIMETER

6·4

Perimeter of a Right Triangle

If you know the lengths of two sides of a **right triangle**, you can find the length of the third side with the **Pythagorean Theorem**.

For a review of the *Pythagorean Theorem,* see page 338.

EXAMPLE **Finding the Perimeter of a Right Triangle**

Use the Pythagorean Theorem to find the perimeter of the right triangle.

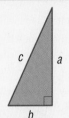

$a = 24$ ft
$b = 10$ ft

$$c^2 = 24^2 + 10^2$$
$$= 576 + 100$$
$$= 676$$
$$\sqrt{c^2} = \sqrt{676}$$
$$c = 26$$

• Use the equation $c^2 = a^2 + b^2$ to find the length of the hypotenuse.

• The square root of c^2 is the length of the hypotenuse.

$24 \text{ ft} + 10 \text{ ft} + 26 \text{ ft} = 60 \text{ ft}$ • Add the lengths of the sides.

So, the perimeter is 60 ft.

Check It Out

Use the Pythagorean Theorem to find the perimeter of each triangle.

8 12 m
15 m

9 12 cm
5 cm

10 17 in.
15 in.

6·4 Exercises

Find the perimeter of each polygon.

1.

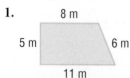

2.

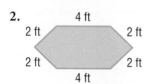

3.

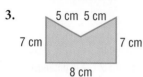

4. Find the perimeter of a square with sides of 5.5 meters.

5. An equilateral triangle has a perimeter of 57 centimeters. Find the length of each side.

6. A regular polygon has sides of 24 centimeters and a perimeter of 360 centimeters. How many sides does the polygon have?

Find the perimeter of each rectangle.

7. $\ell = 10$ in., $w = 7$ in.

8. $\ell = 23$ m, $w = 16$ m

9. $\ell = 6$ cm, $w = 2.5$ cm

10. $\ell = 32$ ft, $w = 1$ ft

Find the perimeter of each triangle.

11.

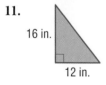

12.

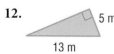

13. Each side of a regular hexagon measures 15 centimeters. What does the hexagon's perimeter measure?

14. The perimeter of a regular octagon is 128 inches. How long is each side?

15. The perimeter of a rectangle is 40 meters. Its width is 8 meters. What is its length?

16. What is the length of each side of a square that has a perimeter of 66 centimeters?

17. Use the Pythagorean Theorem to find the perimeter of △*ABC*.

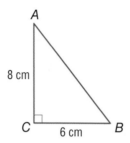

For Exercises 18 and 19, refer to the diagram below.

18. A town is going to put a fence around the park. How long will the fence be?

19. Fencing comes in rolls of 20 meters. How many rolls will be needed?

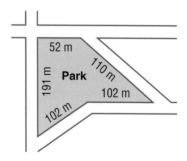

6·5 Area

What Is Area?

Area measures the interior region of a 2-dimensional figure. Instead of measuring with units of length, such as inches, centimeters, feet, and kilometers, area is measured in square units, such as square inches (in^2) and square centimeters (cm^2).

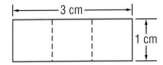

This square has an area of one square centimeter. It takes exactly three of these squares to cover this rectangle, which tells you that the area of the rectangle is three square centimeters, or 3 cm^2.

Estimating Area

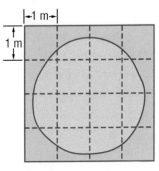

When an exact answer is not needed or is hard to find, you can estimate the area of a surface.

In the shaded figure to the right, four squares are completely shaded, so you know that the area is greater than 4 square meters. The square around the shape covers 16 square meters, and obviously the shaded area is less than that. The estimated area of the shaded figure is greater than 4 square meters but less than 16 square meters. The area is about 8 square meters.

> **Check It Out**
> **Look at the shaded figure.**
> ❶ Estimate the area of the shaded region. Each square represents 1 square meter.

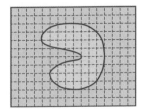

Area of a Rectangle

You can count the squares to find the area of this rectangle.

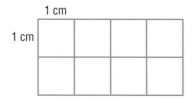

There are 8 squares, and each is a square centimeter. So, the area of this rectangle is 8 square centimeters.

You can also use the formula for finding the area of a rectangle: $A = \ell \times w$. The length of the rectangle above is 4 centimeters and the width is 2 centimeters. Using the formula, you find that

$A = 4$ cm $\times$ 2 cm

$\quad = 8$ cm^2

EXAMPLE Finding the Area of a Rectangle

Find the area of this rectangle.

- The length and the width must be expressed in the same units.

    2 ft = 24 in.

    So, $\ell = 24$ in. and $w = 10$ in.

- Use the formula for the area of a rectangle.

    $A = \ell \times w$

    $\quad = 24$ in. $\times$ 10 in.

    $\quad = 240$ in^2

So, the area of the rectangle is 240 in^2.

If the rectangle is a square, the length and the width are the same. So, for a square whose sides measure s units, you can use the formula $A = s \times s$, or $A = s^2$.

Solve.

2 Find the area of a rectangle with a length of 3 feet and a width of 9 inches.

3 Find the area of a square whose sides measure 13 meters.

Area of a Parallelogram

To find the area of a parallelogram, you multiply the base by the height.

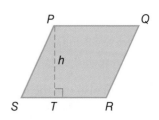

Area = base × height

$A = b \times h$

or $A = bh$

The height of a parallelogram is always **perpendicular** to the base. So, in parallelogram $PQRS$, the height, h, is equal to $\overline{PT}$, not $\overline{PS}$. The base, b, is equal to $\overline{SR}$.

EXAMPLE Finding the Area of a Parallelogram

Find the area of a parallelogram with a base of 9 inches and a height of 5 inches.

$A = b \times h$	• Use the formula for area.
$= 9$ in. $\times$ 5 in.	• Substitute the values for b and h.
$= 45$ in^2	• Solve.

The area of the parallelogram is 45 square inches or 45 in^2.

 Check It Out

Solve.

4 Find the area of a parallelogram with a base of 6 centimeters and a height of 8 centimeters.

5 Find the height of a parallelogram that has a base of 12 feet and an area of 132 square feet.

6 Find the area of a parallelogram with a base of 10 feet and a height of 4 feet.

AREA

6·5

Area of a Triangle

If you were to cut a parallelogram along a diagonal, you would have two triangles with equal bases, b, and the same height, h.

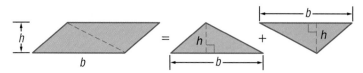

A triangle, therefore, has half the area of a parallelogram with the same base and height. The area of a triangle equals $\frac{1}{2}$ the area of a parallelogram, so the formula is $A = \frac{1}{2} \times b \times h$, or $A = \frac{1}{2}bh$.

$A = \frac{1}{2} \times b \times h$

$A = \frac{1}{2} \times 7.5 \times 8.2$

$\quad = 0.5 \times 7.5 \times 8.2$

$\quad = 30.75 \text{ m}^2$

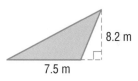

7.5 m

The area of the triangle is 30.75 m².

EXAMPLE **Finding the Area of a Triangle**

Find the area of $\triangle PQR$. Note that in a right triangle, the two **legs** serve as a height and a base.

$A = \frac{1}{2}bh$

$\quad = \frac{1}{2} \times 4 \times 6$

$\quad = 0.5 \times 4 \times 6$

$\quad = 12 \text{ ft}^2$

So, the area of the triangle is 12 ft².
For a review of *right triangles*, see page 338.

Check It Out

Solve.

7 Find the area of a triangle with a base of 15 centimeters and a height of 10 centimeters.

8 Find the area of a right triangle whose sides measure 10 inches, 24 inches, and 26 inches.

Area of a Trapezoid

A trapezoid has two bases, which are labeled b_1 and b_2. You read b_1 as "b sub-one." The area of a trapezoid is equal to the area of two noncongruent triangles.

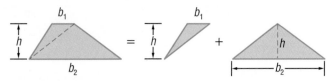

The formula for finding the area of a trapezoid is
$A = \frac{1}{2}b_1 h + \frac{1}{2}b_2 h$ or, in simplified form, $A = \frac{1}{2}h(b_1 + b_2)$.

EXAMPLE **Finding the Area of a Trapezoid**

Find the area of trapezoid $ABCD$.

$A = \frac{1}{2}h(b_1 + b_2)$

$\quad = \frac{1}{2} \times 6\,(8 + 12)$ • Use the formula.

$\quad = 3 \times 20$ • Multiply.

$\quad = 60\ \text{m}^2$ • Solve.

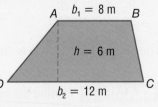

So, the area of the trapezoid is 60 m².

Because $\frac{1}{2}h(b_1 + b_2)$ is equal to $h \times \dfrac{b_1 + b_2}{2}$, you can remember the formula as height times the average of the bases.

For a review of how to find an *average* or *mean,* see page 201.

Check It Out

Solve.

9 The height of a trapezoid is 5 meters. The bases are 3 meters and 7 meters. What is the area?

10 The height of a trapezoid is 2 centimeters. The bases are 8 centimeters and 9 centimeters. What is the area?

11 The height of a trapezoid is 3 inches. The bases are 6 inches and 7 inches. What is the area?

6·5 Exercises

1. Estimate the area of the blue part of the figure below.

2. If each square unit in the figure is 2 square centimeters, estimate the area in square centimeters.

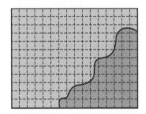

Find the area of each rectangle, with length, ℓ, and width, w.

3. $\ell = 14$ in., $w = 7$ in.

4. $\ell = 19$ cm, $w = 1$ m

Find the area of each parallelogram.

5.

12 ft

6 ft

6.

2.5 m

8 m

Find the area of each triangle, given base, b, and height, h.

7. $b = 16$ cm, $h = 10$ cm

8. $b = 4$ ft, $h = 3.5$ ft

9. A trapezoid has a height of 2 feet. Its bases measure 6 inches and 1 foot. What is its area?

10. Find the area of the figure below.

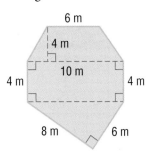

6 m

4 m

10 m

4 m   4 m

8 m   6 m

6•6 Surface Area

The **surface area** of a solid is the total area of its exterior surfaces. You can think about surface area in terms of the parts of a solid shape that you would paint. Like area, surface area is expressed in square units. To see why, "unfold" the rectangular prism.

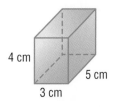

4 cm
5 cm
3 cm

Mathematicians call this unfolded prism a *net*. A net is a pattern that can be folded to make a three-dimensional figure.

Surface Area of a Rectangular Prism

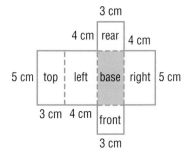

3 cm
4 cm rear 4 cm
5 cm top left base right 5 cm
3 cm 4 cm front
3 cm

A rectangular prism has six rectangular faces. The surface area of a rectangular prism is the sum of the areas of the six faces, or rectangles. For a review of *polyhedrons* and *prisms*, see pages 302–303.

EXAMPLE Finding Surface Area of a Rectangular Prism

Use the net to find the area of the rectangular prism above.

• Use the formula $A = \ell w$ to find the area of each face.

• Add the six areas.

• Express the answer in square units.

Area	=	top + base	+	left + right	+	front + rear
	=	$2 \times (3 \times 5)$	+	$2 \times (4 \times 5)$	+	$2 \times (3 \times 4)$
	=	2×15	+	2×20	+	2×12
	=	30 cm^2	+	40 cm^2	+	24 cm^2
	=	94 cm^2				

So, the surface area of the rectangular prism is 94 cm^2.

Check It Out

Find the surface area of each shape.

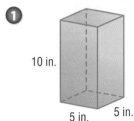
① 10 in.
5 in.   5 in.

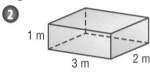
② 1 m
3 m   2 m

Surface Area of Other Solids

Nets can be used to find the surface area of any polyhedron. Look at the **triangular prism** and its net.

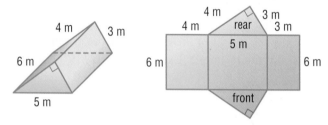

4 m   3 m
6 m
5 m

4 m   3 m
4 m   rear   3 m
5 m
6 m   6 m
front

Use the area formulas for a rectangle ($A = \ell w$) and a triangle ($A = \frac{1}{2}bh$) to find the areas of the five faces. Then find the sum of the areas.

Below are two pyramids and their nets. As with prisms, use the area formulas for a rectangle ($A = \ell w$) and a triangle ($A = \frac{1}{2}bh$). The sum of the areas of the faces is equal to the surface area of the pyramid.

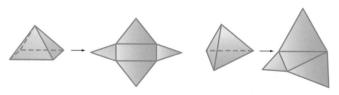

Rectangular pyramid   *Tetrahedron (triangular pyramid)*

The surface of a **cylinder** is the sum of the areas of two circles and a rectangle.

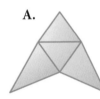
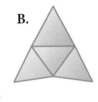

The two bases of a cylinder are equal in area. The height of the rectangle is the height of the cylinder. Its length is the **circumference** of the cylinder.

To find the surface area of a cylinder use the formula for the area of a circle to find the area of each base.

$$A = \pi r^2$$

Then multiply the height of the cylinder by the circumference to find the area of the rectangle. Use the formula $h \times (2\pi r)$.

For a review of *circles,* see page 332.

For a review of *circles,* see page 332.

Check It Out

Solve.

3 Find the surface area of the triangular prism.

4 Which net represents the pyramid?

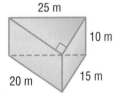

A.   **B.**

5 Find the surface area of the cylinder.
Use $\pi \approx 3.14$.

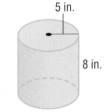

6·6 Exercises

1. Find the surface area of a rectangular prism with sides of 2 meters, 3 meters, and 6 meters.

2. Find the surface area of a rectangular prism with sides of 1 inch, 12 inches, and 12 inches.

3. Find the surface area of a cube with sides of 1 inch.

4. A cube has a surface area of 150 square meters. Find the length of each side.

Find the surface area of each triangular prism.

5.

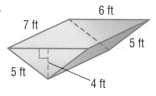

6.

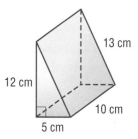

7. Which net shows a regular tetrahedron?

A.

B.

Find the surface area of each cylinder. Use π ≈ 3.14.

8.

9.

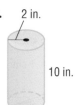

10. Find the surface area of the square pyramid. All the triangular faces are congruent.

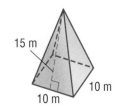

6·7 Volume

What Is Volume?

Volume is the space occupied by a solid. One way to measure volume is to count the number of cubic units that would fill the space inside a figure.

The volume of the small cube is 1 cubic inch.

1 in.
1 in. 1 in.

The number of smaller cubes that it takes to fill the space inside the larger cube is 8, so the volume of the larger cube is 8 cubic inches.

You measure the volume of shapes in *cubic* units. For example, 1 cubic inch is written as 1 in³, and 1 cubic meter is written as 1 m³.

For a review of *cubes,* see page 303.

➡ Check It Out
What is the volume of each shape?

1 1 cube = 1 m³

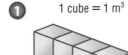

2 1 cube = 1 cm³

3 1 cube = 1 cm³

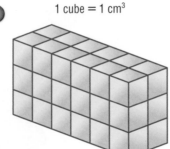

Volume of a Prism

The volume of a prism can be found by multiplying the *area* (pp. 318–322) of the base, *B*, and the height, *h*.

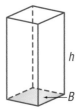 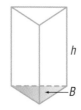 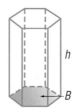

Volume = *Bh*

See *Formulas,* pages 60–61.

EXAMPLE Finding the Volume of a Prism

Find the volume of the rectangular prism.

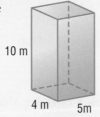

10 m

4 m   5m

base *B* = 5 m × 4 m   • Find the area of the base.

= 20 m²   • Solve.

V = 20 m² × 10 m   • Multiply the base and the height.

= 200 m³   • Solve.

So, the volume of the prism is 200 m³.

 Check It Out

Find the volume of each shape.

④
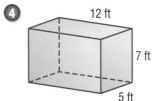
12 ft

7 ft

5 ft

⑤

6 in.

4 in.   4 in.

Volume of a Cylinder

You can find the volume of a cylinder the same way you found
the volume of a prism, using the formula $V = Bh$. *Remember:*
The base of a cylinder is a circle.

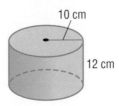

You can use the formula $V = Bh$. The base has a radius of
10 centimeters, so the area (πr^2) is $\pi \times 100$ square centimeters,
about 314 square centimeters. Then multiply the area of the base
by the height.

$$V \approx 314 \text{ cm}^2 \times 12 \text{ cm}$$
$$\approx 3{,}768 \text{ cm}^3$$

The volume of the cylinder is 3,768 cm³.

Check It Out
Find the volume of each cylinder. Use $\pi \approx 3.14$.

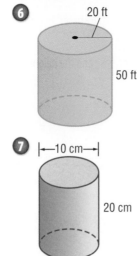

6 20 ft, 50 ft

7 |—10 cm—|, 20 cm

VOLUME

6•7

6·7 Exercises

1. A rectangular prism has sides of 4 inches, 9 inches, and 12 inches. Find the volume of the prism.

2. The volume of a rectangular prism is 140 cubic feet. The length of the base is 7 feet and the width of the base is 4 feet. What is the height of the prism?

3. Find the volume of a cube with sides of 10 centimeters.

4. A cube has a volume of 125 cubic meters. What is the length of each side of the cube?

Find the volume of each solid. Use π ≈ 3.14.

5.
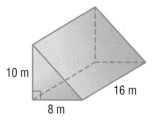
10 m
8 m
16 m

6. 2 cm
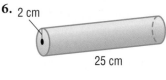
25 cm

7. |←—12 in.—→|
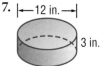
3 in.

8.
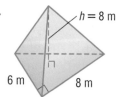
h = 8 m
6 m
8 m

9.
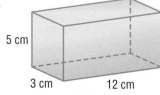
5 cm
3 cm
12 cm

10.
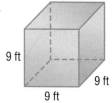
9 ft
9 ft
9 ft

6·8 Circles

Parts of a Circle

Circles differ from other geometric shapes in several ways. For instance, all circles are the same shape; polygons vary in shape. Circles have no sides, but polygons are named and classified by the number of sides they have. The *only* thing that makes one circle different from another is size.

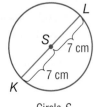

Circle *S*

A circle is a set of points equidistant from a given point. That point is the center of the circle. A circle is named by its center point.

A **radius** is any **segment** that has one endpoint at the center and the other endpoint on the circle. In circle *S*, $\overline{SL}$ is a *radius,* and so is $\overline{SK}$.

A **diameter** is any line segment that passes through the center of the circle and has both endpoints on the circle. $\overline{KL}$ is a diameter of circle *S*. The length of the diameter $\overline{KL}$ is equal to the sum of radii $\overline{SK}$ and $\overline{SL}$. In other words, the diameter, *d,* is twice the length of the radius, *r*: $d = 2r$. The diameter of circle *S* is 2(7) or 14 centimeters.

Check It Out
Solve.

1. Find the radius of a circle with diameter 26 meters.
2. Find the radius of a circle with diameter 1 centimeter.
3. Find the diameter of a circle in which $r = 16$ inches.
4. Find the diameter of a circle in which $r = 2.5$ feet.
5. The diameter of circle *P* measures twice the diameter of circle *Q*. The radius of circle *Q* measures 6 meters. What is the length of the radius of circle *P*?

Circumference

The circumference of a circle C is the distance around the circle. The ratio of circumference, C, to diameter, d, is always the same. This ratio is a number close to 3.14. In other words, the circumference is about 3.14 times the diameter. The symbol π, which is read as **pi**, is used to represent the ratio $\frac{C}{d}$.

$$\frac{C}{d} \approx 3.141592\ldots$$

Circumference = pi × diameter, or $C = \pi d$

You can find the diameter of a circle if you know its circumference. $d = \frac{C}{\pi}$

Because $d = 2r$, we can also define circumference in terms of the radius.

Circumference = 2 × pi × radius or $C = 2\pi r$.

Look at the illustration below. The circumference of the circle is a little longer than the length of three diameters. This is true for any circle.

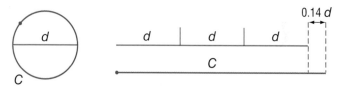

A calculator that has a π key will give an approximation for π to several decimal places: $\pi \approx 3.141592\ldots$. For practical purposes, however, π is often rounded to 3.14 or simply left as π.

EXAMPLE Finding the Circumference of a Circle

Find the circumference, to the nearest tenth, of a circle with radius 6 meters.

$d = 6 \text{ m} \times 2 = 12 \text{ m}$	• To find the diameter, multiply the radius by 2.
$C = \pi \times 12 \text{ m}$	• Use the formula $C = \pi d$.
The exact circumference is 12π m.	• You can leave the answer in terms of π.
$C \approx 3.14 \times 12 \text{ m}$	• Or, use $\pi \approx 3.14$ to estimate.
$\approx 37.68 \text{ m} \approx 37.7 \text{ m}$	• Round your answer, if necessary.

So, to the nearest tenth, the circumference is 37.7 meters.

Check It Out

Give your answers in terms of π.

6 What is the circumference of a circle with radius 14 feet?

7 What is the circumference of a circle with diameter 21 centimeters?

Use $\pi \approx 3.14$. Round your answers to the nearest tenth.

8 Find the circumference of a circle with diameter 14.6 meters.

9 Find the circumference of a circle with radius 18 centimeters.

10 Find the diameter of a circle that has a circumference of 20.41 inches.

Area of a Circle

To find the area of a circle, you use the formula: $A = \pi r^2$.
As with the area of polygons, the area of a circle is expressed
in square units.

For a review of *area* and *square units,* see page 318.

EXAMPLE Finding the Area of a Circle

Find the area of circle T to the nearest whole number.

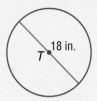

$r = 18$ in. $\div 2 = 9$ in.	• To find the radius, divide the diameter by 2.
$A = \pi \times (9)^2 = 81\pi$ in^2	• Use the formula $A = \pi r^2$.
≈ 254.34 in^2	• Use $\pi \approx 3.14$.
≈ 254 in^2	• Round to the nearest whole number.

So, the area of circle T is about 254 in^2.

 Check It Out

Find the area.

11 Find the area of a circle with radius 13 centimeters.
Give your answer in terms of π.

12 A circle has a diameter of 21 feet. Find the area of the
circle to the nearest whole number. Use $\pi \approx 3.14$.

13 Find the area of a circle with a diameter of 16 centimeters.
Give your answer in terms of π.

Now, That's a Pizza!

Put together 10,000 pounds of flour, 664 gallons of water, 316 gallons of tomato sauce, 1,320 pounds of cheese, and 1,200 pounds of pepperoni. What have you got? An 18,664-pound pizza, baked by L. Amato in 1978 to raise money for charity.

Together with L. Piancone, Amato organized the baking of another giant pizza in 1991. This pie, with a diameter of about $56\frac{1}{2}$ feet, still holds the record for the largest pizza made in the United States. What was its area? See **HotSolutions** for the answer.

6·8 Exercises

1. What is the radius of a circle with diameter 42.8 meters?

2. The diameter of a circle is 10 centimeters greater than the radius. How long is the radius?

Find the circumference to the nearest tenth of each circle with given radius or diameter. Use $\pi \approx 3.14$.

3. $d = 15$ in.

4. $d = 7$ m

5. $r = 5.5$ cm

6. The circumference of a circle measures 63.4 feet. Find the circle's diameter to the nearest tenth.

7. Find the radius to the nearest whole number of a circle that has a circumference of 1,298 meters.

Find the area in terms of π of each circle with given radius or diameter.

8. $r = 11$ ft

9. $d = 60$ cm

10. $r = 1.5$ in.

Find the area to the nearest tenth of each circle with given radius or diameter. Use $\pi \approx 3.14$.

11. $d = 16$ m

12. $r = 9$ ft

13. $r = 12.8$ cm

14. A circle has a circumference of 25 inches. Find the area of the circle to the nearest whole number.

15. If you double the diameter of a circle, you increase its circumference by ____.
    A. 2 times   **B.** π times   **C.** 2^2 times

16. If you double the radius of a circle, you increase its area by ____.
    A. 2 times   **B.** π times   **C.** 2^2 times

6·9 Pythagorean Theorem

Right Triangles

The larger illustration at the right shows a right triangle on a geoboard. You can count that leg a is 2 units long and leg b is 3 units long. The **hypotenuse**, side c, is always opposite the right angle. You cannot count its length.

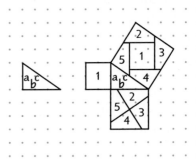

Now look at the smaller illustration on the left. A square is drawn on each of the three sides of the triangle.

The first square is placed on leg a and labeled 1. Using the formula for the area of a square $(A = s^2)$, the area of the first square is $A = 2^2 = 4$ square units.

The second square is placed on leg b. It is made of four congruent quadrilaterals labeled 2, 3, 4, and 5. The area of the second square is $A = 3^2 = 9$ square units.

The third square is placed on leg c. Both the first square and the second square can be combined to make the third square. The area of the third square must be equal to the sum of the areas of the first and second squares. The area of the third square is $A = 2^2 + 3^2 = 4 + 9 = 13$ square units.

Note the relationship among the three areas:
   area of square 1 + area of square 2 = area of square 3
Therefore, $a^2 + b^2 = c^2$.

This relationship holds true for all right triangles.

1 What is the area of each of the squares?

2 What is the relationship among the squares?

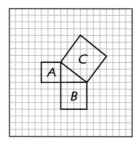

The Pythagorean Theorem

The Pythagorean Theorem can be stated as follows: In a right triangle, the square of the length of the hypotenuse is equal to the sum of the squares of the lengths of the legs.

$$c^2 = a^2 + b^2$$

You can use the Pythagorean Theorem to find the length of the third side of a right triangle if you know the lengths of the other two sides.

EXAMPLE	Using the Pythagorean Theorem to Find the Hypotenuse

Use the Pythagorean Theorem, $c^2 = a^2 + b^2$, to find the length of the hypotenuse, c, of $\triangle WXY$.

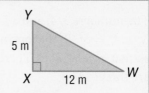

$c^2 = 5^2 + 12^2$ • Substitute known lengths for a and b.

$c^2 = 25 + 144$ • Square the two known lengths.

$c^2 = 169$ • Find the sum of the squares of the two legs.

$c = 13$ • Take the square root of the sum.

So, the hypotenuse measures 13 meters.

EXAMPLE | Using the Pythagorean Theorem to Find a Side Length

Use the Pythagorean Theorem, $c^2 = a^2 + b^2$, to find the length of the unknown leg, b, of a right triangle with a hypotenuse of 10 centimeters and one leg measuring 4 centimeters. Round to the nearest tenth of a centimeter.

$10^2 = 4^2 + b^2$	• Use $c^2 = a^2 + b^2$.
$100 = 16 + b^2$	• Square the known lengths.
$100 - 16 = (16 - 16) + b^2$	• Subtract to isolate unknown.
$84 = b^2$	
$9.16515 \ldots = b$	• Use your calculator to find the square root.

So, the length of the unknown side is about 9.2 centimeters.

 Check It Out

Solve.

③ Find the length to the nearest whole number of the hypotenuse of a right triangle with legs measuring 5 meters and 6 meters.

④ Find the length of $\overline{KL}$ to the nearest whole number.

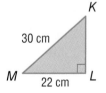

Pythagorean Triples

The numbers 3, 4, and 5 form a **Pythagorean triple** because $3^2 + 4^2 = 5^2$. Pythagorean triples are formed by whole numbers, so that $a^2 + b^2 = c^2$. There are many Pythagorean triples. Here are three:

$$5, 12, 13 \qquad 8, 15, 17 \qquad 7, 24, 25$$

If you multiply each number of a Pythagorean triple by the same number, you form another Pythagorean triple. 6, 8, 10 is a triple because it is 2(3), 2(4), 2(5).

6•9 Exercises

1. What is the relationship among the lengths *x*, *y*, and *z*?

Find the missing length in each right triangle. Round to the nearest tenth.

2.

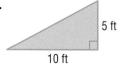

5 ft
10 ft

3.

12 cm   14 cm

4.

7 in.
7 in.

5.

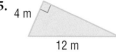

4 m
12 m

Are the following sets of numbers Pythagorean triples? Write *yes* or *no*.

6. 9, 12, 15   **7.** 10, 24, 26   **8.** 8, 16, 20

Use the Pythagorean Theorem to find the missing lengths.

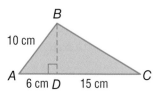

B
10 cm
A   6 cm D   15 cm   C

9. Find the length of $\overline{BD}$.

10. Find the length of $\overline{BC}$.

Geometry

What have you learned?

You can use the problems and the list of words that follow to see what you learned in this chapter. You can find out more about a particular problem or word by referring to the topic number (*for example,* Lesson 6·2).

Problem Set

Use the figure to answer Exercises 1–3.

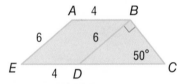

1. Name three obtuse angles in the figure. (Lesson 6·1)
2. Find $m\angle BDC$. (Lesson 6·1)
3. What kind of quadrilateral is *ABDE*? (Lesson 6·2)

4. Write the letters of your first and last names. Do any letters have more than one line of symmetry? If so, which ones?
   (Lesson 6·3)

5. Find the area of a triangle with base 9 feet and height 14 feet.
   (Lesson 6·5)

6. A cube has a volume of 64 cubic inches. Find the length of the sides of the cube. (Lesson 6·7)

Use the figure to answer Exercise 7.

7. What is the circumference of circle *S* in terms of π? (Lesson 6·8)

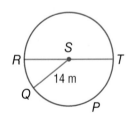

Use the figure to answer Exercises 8 and 9.

8. Find the length of $\overline{KM}$. (Lesson 6·9)

9. Find the length of $\overline{KN}$. (Lesson 6·9)

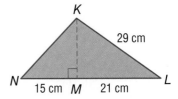

K

29 cm

N   15 cm   *M*   21 cm   *L*

HotWords

Write definitions for the following words.

acute angle (Lesson 6·1)

angle (Lesson 6·1)

circumference (Lesson 6·6)

complementary angle (Lesson 6·1)

congruent angle (Lesson 6·1)

cube (Lesson 6·2)

cylinder (Lesson 6·6)

degree (Lesson 6·1)

diagonal (Lesson 6·2)

diameter (Lesson 6·8)

face (Lesson 6·2)

hypotenuse (Lesson 6·9)

isosceles triangle (Lesson 6·1)

legs of a triangle (Lesson 6·5)

line of symmetry (Lesson 6·3)

opposite angle (Lesson 6·2)

parallelogram (Lesson 6·2)

perimeter (Lesson 6·4)

perpendicular (Lesson 6·5)

pi (Lesson 6·8)

point (Lesson 6·1)

polygon (Lesson 6·1)

polyhedron (Lesson 6·2)

prism (Lesson 6·2)

Pythagorean Theorem (Lesson 6·4)

Pythagorean triple (Lesson 6·9)

quadrilateral (Lesson 6·2)

radius (Lesson 6·8)

ray (Lesson 6·1)

rectangular prism (Lesson 6·2)

reflection (Lesson 6·3)

regular polygon (Lesson 6·2)

regular shape (Lesson 6·2)

right angle (Lesson 6·1)

right triangle (Lesson 6·4)

rotation (Lesson 6·3)

segment (Lesson 6·8)

supplementary angle (Lesson 6·1)

surface area (Lesson 6·6)

tetrahedron (Lesson 6·2)

transformation (Lesson 6·3)

translation (Lesson 6·3)

trapezoid (Lesson 6·2)

triangular prism (Lesson 6·6)

vertex (Lesson 6·1)

vertical angle (Lesson 6·1)

volume (Lesson 6·7)

HotTopic 7

Measurement

What do you know?

You can use the problems and the list of words that follow to see what you already know about this chapter. The answers to the problems are in **HotSolutions** at the back of the book, and the definitions of the words are in **HotWords** at the front of the book. You can find out more about a particular problem or word by referring to the topic number (*for example*, Lesson 7·2).

Problem Set

1. Name one metric unit and one customary unit of weight.
   (Lesson 7·1)

2. Name one metric unit and one customary unit of distance.
   (Lesson 7·1)

Complete the conversions. (Lessons 7·2 and 7·3)

3. $4.5 \text{ km} = \underline{\hphantom{xx}} \text{ m}$

4. $8.4 \text{ m} = \underline{\hphantom{xx}} \text{ mm}$

5. $4 \text{ yd} = \underline{\hphantom{xx}} \text{ in.}$

6. $1{,}320 \text{ ft} = \underline{\hphantom{xx}} \text{ mi}$

7. $23.8 \text{ cm}^2 = \underline{\hphantom{xx}} \text{ mm}^2$

8. $7 \text{ yd}^2 = \underline{\hphantom{xx}} \text{ ft}^2$

9. $1 \text{ yd}^2 = \underline{\hphantom{xx}} \text{ in}^2$

10. $1.9 \text{ m}^3 = \underline{\hphantom{xx}} \text{ cm}^3$

Use the rectangle to answer Exercises 11 and 12.

12 ft

3 ft

11. What is the rectangle's perimeter in inches? in yards? (Lesson 7·2)

12. What is the rectangle's area in square inches? (Lesson 7·3)

Use the rectangular prism to answer Exercises 13–15. (Lesson 7·3)

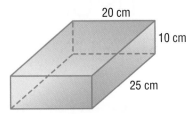

13. What is the prism's volume in cubic millimeters?
14. What is the prism's volume in cubic meters?
15. If 1 inch = 2.54 centimeters, what is the approximate volume of the prism in cubic inches?

16. A recipe calls for 10 ounces of rice for 4 people. How many pounds of rice do you need to make the recipe for 24 people? (Lesson 7·4)

17. How many 200-gram packets of dried beans would you have to buy if you needed 2.4 kilograms of beans? (Lesson 7·4)

A photograph that was 9 inches long by 12 inches high was reduced so that the new photo is 8 inches high. (Lesson 7·5)

18. What is the new length of the photo?
19. What is the ratio of the area of the new photo to the area of the original photo?

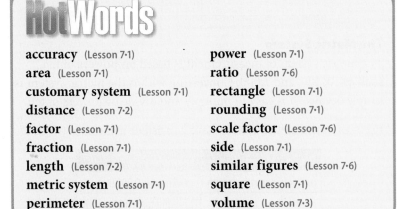

HotWords

accuracy (Lesson 7·1)	**power** (Lesson 7·1)
area (Lesson 7·1)	**ratio** (Lesson 7·6)
customary system (Lesson 7·1)	**rectangle** (Lesson 7·1)
distance (Lesson 7·2)	**rounding** (Lesson 7·1)
factor (Lesson 7·1)	**scale factor** (Lesson 7·6)
fraction (Lesson 7·1)	**side** (Lesson 7·1)
length (Lesson 7·2)	**similar figures** (Lesson 7·6)
metric system (Lesson 7·1)	**square** (Lesson 7·1)
perimeter (Lesson 7·1)	**volume** (Lesson 7·3)

7·1 Systems of Measurement

The Metric and Customary Systems

If you watch the Olympic Games, you may notice that the distances are measured in meters or kilometers, and weights are measured in kilograms. That is because the most common system of measurement in the world is the **metric system**. The **customary system** of measurement is used in the United States. It may be useful to make conversions from one unit of measurement to another within each system, as well as convert units between the two systems.

Basic Measures			
	Metric		**Customary**
Distance:	meter		inch, foot, yard, mile
Capacity:	liter		cup, quart, gallon
Weight:	gram		ounce, pound, ton

The Metric System

The metric system of measurement is based on **powers** of ten, such as 10, 100, and 1,000. Converting within the metric system is simple because it is easy to multiply and divide by powers of ten.

Prefixes in the metric system have consistent meanings.

Prefix	Meaning	Example
milli-	one thousandth	1 *milli*liter is 0.001 liter.
centi-	one hundredth	1 *centi*meter is 0.01 meter.
kilo-	one thousand	1 *kilo*gram is 1,000 grams.

The Customary System

The customary system of measurement is not based on powers of ten. It is based on numbers like 12 and 16, which have many factors. This makes it easy to find, say, $\frac{2}{3}$ ft or $\frac{3}{4}$ lb. While the metric system uses decimals, you will frequently encounter fractions in the customary system.

Unfortunately, there are no convenient prefixes in the customary system, so you need to memorize the basic equivalent units.

Customary Units		
Type of Measure	**Larger Unit** ⟶	**Smaller Units**
Length	1 foot (ft) =	12 inches (in.)
	1 yard (yd) =	3 feet
	1 mile (mi) =	5,280 feet
Weight	1 pound (lb) =	16 ounces (oz)
	1 ton (T) =	2,000 pounds
Capacity	1 cup (c) =	8 fluid ounces (fl oz)
	1 pint (pt) =	2 cups
	1 quart (qt) =	2 pints
	1 gallon (gal) =	4 quarts

 Check It Out

Identify the system of measurement.

1. Which system is based on multiples of 10?
2. Which system uses fractions?

7·1 Exercises

1. What are the metric prefixes and their meaning?
2. How many meters are in a kilometer?
3. How many centimeters are in a kilometer?
4. How many millimeters are in 1 centimeter?

Which system of measurement uses the following?

5. ounces
6. liters
7. kilograms
8. yards

Which is the larger unit of measurement?

9. pound or ounce
10. quart or gallon

Which is the smaller unit of measurement?

11. yard or foot
12. cup or pint

Identify the equivalent units in the customary system.

13. 1 foot = ____ inches
14. 1 mile = ____ feet
15. 16 ounces = ____ pound(s)
16. 4 quarts = ____ gallon(s)
17. 1 cup = ____ fluid ounces
18. 1 yard = ____ feet
19. 2 pints = ____ quart(s)
20. 1 ton = ____ pound(s)

7·2 Length and Distance

About What Length?

When you get a feel for "about how long" or "about how far," it is easier to make estimations about length and distance. Here are some everyday items that will help you keep in mind what metric and customary units mean.

Metric Units	Customary Units
millimeter	**inch**
1 mm	1 in.
centimeter	**foot**
1 cm	1 ft
meter	**yard**
1 m	1 yd

Metric and Customary Units

When you are measuring **length** or **distance**, you may use two different systems of measurement. One is the metric system, and the other is the customary system.

Metric Equivalents						
1 km	=	1,000 m	=	100,000 cm	=	1,000,000 mm
0.001 km	=	1 m	=	100 cm	=	1,000 mm
		0.01 m	=	1 cm	=	10 mm
		0.001 m	=	0.1 cm	=	1 mm

Customary Equivalents						
1 mi	=	1,760 yd	=	5,280 ft	=	63,360 in.
$\frac{1}{1,760}$ mi	=	1 yd	=	3 ft	=	36 in.
		$\frac{1}{3}$ yd	=	1 ft	=	12 in.
		$\frac{1}{36}$ yd	=	$\frac{1}{12}$ ft	=	1 in.

EXAMPLE   Changing Units Within a System

How many feet are in $\frac{1}{8}$ mile?

units you have

1 mi = 5,280 ft.

conversion factor for new units

$\frac{1}{8} \times 5,280 = 660$

There are 660 feet in $\frac{1}{8}$ mile.

- Use the equivalents chart to find where the units you have equal 1. That is, how many feet are in 1 mile?

- Find the conversion factor.

- Multiply to complete the conversion.

Check It Out

Identify the equivalent unit.

1 2,200 m = ____ km

2 60 in. = ____ ft

3 3 yd = ____ ft

4 $\frac{1}{2}$ km = ____ m

Conversions Between Systems

Sometimes, you may want to convert between the metric system and the customary system. You can use this conversion table to help.

Conversion Table				
1 inch	=	25.4 millimeters	1 millimeter =	0.0394 inch
1 inch	=	2.54 centimeters	1 centimeter =	0.3937 inch
1 foot	=	0.3048 meter	1 meter =	3.2808 feet
1 yard	=	0.914 meter	1 meter =	1.0936 yards
1 mile	=	1.609 kilometers	1 kilometer =	0.621 mile

To make a conversion, find the listing where the unit you have is 1. Multiply the number of units you have by the conversion factor for the new units.

EXAMPLE Changing Units Between Systems

127 cm = _____ in.

1 cm = 0.3937 in. • Find the conversion factor.

127 × 0.3937 • Multiply.

≈ 50 • Solve.

So, 126 cm = about 50 in.

Most of the time you just need to estimate the conversion from one system to the other to get an idea of the size of your item. Round off numbers in the conversion table to simplify your thinking. Think that 1 meter is just a little more than 1 yard, 1 inch is between 2 and 3 centimeters, 1 mile is about $1\frac{1}{2}$ kilometers. When your friend in Canada says she caught a fish 60 cm long, you know that it is between 20 and 30 in. long.

Check It Out

Use a calculator and round to the nearest tenth.

5 Change 31 in. to cm.   **6** Change 64 m to yd.

7·2 Exercises

Complete the conversions.

1. 1.93 km = ____ m

2. 45 cm = ____ mm

3. 750 cm = ____ m

4. 820 m = ____ km

5. 252 in. = ____ ft

6. 10 yd = ____ in.

7. 440 yd = ____ mi

8. 1.5 mi = ____ ft

9. 396 in. = ____ yd

10. 12.5 ft = ____ in.

Use a calculator and round to the nearest tenth.

11. Change 2 ft to in.

12. Change 5.4 m to cm.

13. Change 14 mi to km.

14. Change 420 mm to in.

15. Change 32 km to mi.

16. Change 15 yd to m.

Choose the conversion estimate.

17. 550 yd is about ____.
    - **A.** 0.5 km
    - **B.** 5 km
    - **C.** 50 m

18. 1 m is about ____.
    - **A.** 1 ft
    - **B.** 10 ft
    - **C.** 3 ft

19. 5 mi is about ____.
    - **A.** 10 km
    - **B.** 3 km
    - **C.** 8 km

20. 10 in. is about ____.
    - **A.** 25 cm
    - **B.** 4 cm
    - **C.** 250 cm

7·3 Area, Volume, and Capacity

Area

Area is the measure of a surface in square units. The large surface of the United States covers a land area of 3,787,319 square miles. Area can be measured in metric units or customary units. Sometimes you convert measurements within a measurement system (p. 350). Below is a chart that provides some of the most common conversions.

Metric	Customary
$100 \text{ mm}^2 = 1 \text{ cm}^2$	$144 \text{ in}^2 = 1 \text{ ft}^2$
$10,000 \text{ cm}^2 = 1 \text{ m}^2$	$9 \text{ ft}^2 = 1 \text{ yd}^2$
	$4,840 \text{ yd}^2 = 1 \text{ acre}$
	$640 \text{ acres} = 1 \text{ mi}^2$

To convert to a new unit, find the listing where the unit you have is equal to one. Multiply the number of units you have by the conversion factor for the new unit.

EXAMPLE Converting Within a Measurement System

If the United States covers an area of about 3,800,000 mi², how many acres does it cover?

$3,800,000 \text{ mi}^2 = $ _____ acres

$1 \text{ mi}^2 = 640$ acres • Find the conversion factor.

$3,800,000 \times 640 = 2,432,000,000$ • Calculate the conversion.

So, the United States covers 2,432,000,000 acres.

 Check It Out

Solve.

❶ $7 \text{ yd}^2 = $ _____ ft^2 ❷ $4 \text{ m}^2 = $ _____ cm^2

Volume

Volume is expressed in cubic units. Here are the basic relationships among units of volume.

Metric	Customary
$1,000 \text{ mm}^3 = 1 \text{ cm}^3$	$1,728 \text{ in}^3 = 1 \text{ ft}^3$
$1,000,000 \text{ cm}^3 = 1 \text{ m}^3$	$27 \text{ ft}^3 = 1 \text{ yd}^3$

EXAMPLE Converting Volume Within a System of Measurement

Express the volume of the carton in cubic feet.

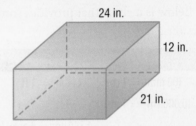

24 in.

12 in.

21 in.

$V = \ell wh$
  $= 24 \times 21 \times 12$
  $= 6,048 \text{ in}^3$

$1,728 \text{ in}^3 = 1 \text{ ft}^3$

$6,048 \div 1,728 = 3.5 \text{ ft}^3$

• Use a formula to find the volume (p. 61), using the units of the dimensions.

• Find the conversion factor.

• Multiply to convert to smaller units. Divide to convert to larger units. Include the unit of measurement in your answer.

So, the volume of the carton is 3.5 ft^3.

Check It Out
Find the volume.

❸ Find the volume of a cube with sides of 50 centimeters. Convert your answer to cubic meters.

❹ Find the volume of a rectangular prism with sides of 5 feet, 3 feet, and 18 feet. Convert your answer to cubic yards.

Capacity

Capacity is closely related to volume, but there is a difference. A block of wood has volume but no capacity to hold liquid. The capacity of a container is a measure of the volume of liquid it will hold.

Metric	Customary
1 liter (L) = 1,000 milliliters (mL)	8 fl oz = 1 cup (c)
	2 c = 1 pint (pt)
	2 pt = 1 quart (qt)
	4 qt = 1 gallon (gal)
Conversion Factor: 1 L = 1.057 qt	

Note the use of *fl oz* (fluid ounce) in the table. This is to distinguish it from *oz* (ounce), which is a unit of weight (16 oz = 1 lb.) Fluid ounce is a unit of capacity (16 fl oz = 1 pint). There is a connection between ounce and fluid ounce. A pint of water weighs about a pound, so a fluid ounce of water weighs about an ounce. For water, as well as for most other liquids, *fluid ounce* and *ounce* are equivalent, and the "fl" is sometimes omitted (for example, "8 oz = 1 cup"). To be correct, though, use *ounce* for weight only and *fluid ounce* for capacity. For liquids that weigh considerably more or less than water, the difference is significant.

A gallon of orange juice costs $5.51. How much does it cost per liter?

> Figure out how many liters are in a gallon. There are 4 quarts in a gallon, so there are 4 × 1.057 liters, or 4.228 liters, in a gallon. So, 1 liter of orange juice costs $5.51 ÷ 4.228 or $1.30.

Check It Out

Identify the better buy.

5 1 liter of milk for $1.87 or 1 quart of milk for $1.22

6 1 pint of orange juice for $2.28 or 1 liter for $3.41

7 1 liter of gasoline for $3.98 or 1 gallon for $3.78

7·3 Exercises

Identify each unit as a measure of distance, area, volume, or capacity.

1. liter   **2.** km^3

3. mm   **4.** mi^2

Give the area of the rectangle in each of the units.

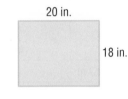

20 in.

18 in.

5. in^2 **6.** ft^2

Give the volume of the rectangular prism in each of the units.

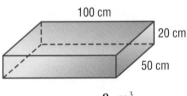

100 cm

20 cm

50 cm

7. cm^3 **8.** m^3

9. Find the area of a square with sides of 4.5 ft. Give the answer in square yards.

10. Find the volume of a cube with sides of 150 cm. Give the answer in cubic meters.

Complete the conversions.

11. 20 pt = ____ gal   **12.** 12 c = ____ qt

13. 1 qt = ____ fl oz   **14.** 3.5 gal = ____ c

15. 96 fl oz = ____ gal   **16.** 2.75 gal = ____ pt

17. 400 mL = ____ L   **18.** 3 L = ____ mL

19. 1,750 mL = ____ L

20. Box A has a volume of 140 cubic feet, and Box B has a volume of 5 cubic yards. Which is the larger box?

7·4 Mass and Weight

Though mass and weight are similar, they are not the same.

Mass is the measure
of the amount of
matter in an object.

Weight is the measure
of how heavy
an object is.

Gravity influences weight but does not affect mass. Your weight would change if you were on the Moon. Your mass would not change.

Customary and Metric Relationships

To convert from one unit of mass or weight to another, use the conversions below.

Customary and Metric Conversions			
Type of Measure	**Customary**		**Metric**
Length	1 inch (in.)	≈	2.54 centimeters (cm)
	1 foot (ft)	≈	0.30 meter (m)
	1 yard (yd)	≈	0.91 meter (m)
	1 mile (mi)	≈	1.61 kilometers (km)
Weight/Mass	1 pound (lb)	≈	453.6 grams (g)
	1 pound (lb)	≈	0.4536 kilogram (kg)
	1 ton (T)	≈	907.2 kilograms (kg)
Capacity	1 cup (c)	≈	236.59 milliliters (mL)
	1 pint (pt)	≈	473.18 milliliters (mL)
	1 quart (qt)	≈	946.35 milliliters (mL)
	1 gallon (gal)	≈	3.79 liters (L)

7·4 Exercises

Complete the conversions.

1. 40 oz = _____ lb
2. 3,500 lb = _____ T
3. 12.35 g = _____ mg
4. 6,040 mg = _____ g
5. 4.5 lb = _____ oz
6. 85 g = _____ kg
7. 3,200 oz = _____ T
8. 150,000 mg = _____ kg
9. 500 oz = _____ lb
10. 45 mg = _____ g
11. 0.02 kg = _____ mg
12. 0.5 T = _____ oz
13. 1 T = _____ kg
14. 100 g = _____ oz
15. 10 kg = _____ lb
16. 8 oz = _____ g
17. 15 lb = _____ kg

18. A recipe calls for 18 ounces of flour for 6 people. How many pounds of flour do you need to make the recipe for 24 people?

19. A 2-pound bag of dried fruit costs $6.40. A 5-ounce box of the same fruit costs $1.20. Which is the better deal?

20. How many 4-ounce balls of wool would you have to buy if you needed 2 kilograms of wool?

APPLICATION **Poor SID**

SID is a crash-test dummy. After a crash, SID goes to the laboratory for a readjustment of sensors and perhaps a replacement head or other body parts. Because of the forces at work during a car crash, body parts weigh as much as 20 times their normal weight.

The weight of a body changes during a crash. Does the mass of the body also change?

See **HotSolutions** for the answer.

7·5 Size and Scale

Similar Figures

Similar figures are figures that have the same shape but are not necessarily the same size.

EXAMPLE Deciding Whether Two Figures Are Similar

Are these two rectangles similar?

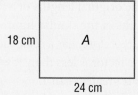

$\dfrac{24}{20} \overset{?}{=} \dfrac{18}{15}$ • Set up using the proportion. $\dfrac{\text{length } A}{\text{length } B} \overset{?}{=} \dfrac{\text{width } A}{\text{width } B}$

$15 \times 24 \overset{?}{=} 18 \times 20$ • Cross multiply to see if ratios are equal.

$360 = 360$ • If all sides have equal ratios, the figures are similar.

So, the rectangles are similar.

Check It Out

① Identify the similar figures.

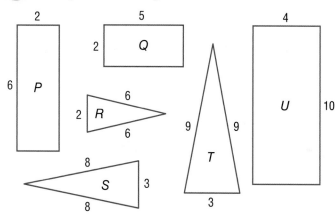

Scale Factors

A **scale factor** is the ratio of
two corresponding sides of
two similar geometric figures.

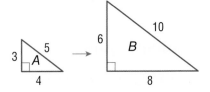

$\triangle A$ is similar to $\triangle B$. The ratio of a side length of $\triangle B$ to the
corresponding side length of $\triangle A$ is $\frac{10}{5}$, or 2. Therefore, the scale
factor is 2.

Enlargements or reductions are expressed in terms of scale factors
rather than ratios. You multiply the original lengths of a figure
by the scale factor to find the lengths of the similar figure. An
exact full-size copy of something, therefore, has a scale factor of 1.
When a figure is reduced, the scale factor is less than 1; when a
figure is enlarged, the scale factor is greater than 1.

EXAMPLE   **Finding the Scale Factor**

What is the scale factor for these similar rectangles?

6

4   C

3

D   2

$\dfrac{D}{C} = \dfrac{3}{6}$ • Set up a ratio of corresponding sides: $\dfrac{\text{new figure}}{\text{original figure}}$

$\quad\;\; = \dfrac{1}{2}$ • Simplify, if possible.

The scale factor of the similar rectangles is $\frac{1}{2}$.

Check It Out

Find the scale factors.

2

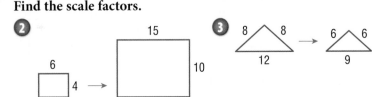

15

6

4   →

10

3   8   8

12

6   6

9

Scale Factors and Area

Scale factors of similar figures can be used to find the ratio of the areas.

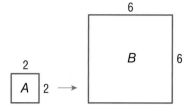

For the squares, the scale factor is 3 because the ratio of sides is $\frac{6}{2} = 3$. Notice that, while the scale factor is 3, the ratio of the areas is 9.

$$\frac{\text{Area of } B}{\text{Area of } A} = \frac{6^2}{2^2} = \frac{36}{4} = 9$$

In general, the ratio of the areas of two similar figures is the *square* of the scale factor.

EXAMPLE Using Scale Factor to Find Area

The scale factor is $\frac{1}{2}$. What is the ratio of the areas?

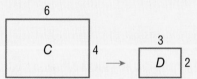

$\dfrac{3 \times 2}{6 \times 4} = \dfrac{6}{24}$ • Set up the ratios. $\dfrac{\text{Area of } D}{\text{Area of } C}$

$\dfrac{6}{24} = \dfrac{1}{4}$ • Solve.

The ratio of the areas is $\left(\frac{1}{2}\right)^2 = \frac{1}{4}$.

 Check It Out

Solve.

4 The scale factor for two similar figures is $\frac{5}{4}$. What is the ratio of the areas?

5 The scale of a house model is 1 foot = 10 feet. How much area of the house does an area of 1 square foot on the model represent?

Scale Factor and Volume

Two solids are similar if all of their dimensions are similar with the same scale factor. Scale factor for volume refers to the ratio or scale factor of the dimensions of two solids.

The ratio of the volume of two solids is the cube of the ratio given by the scale factor.

EXAMPLE Using Scale Factor to Find Volume

The scale factor for two similar solids is $\frac{1}{8}$. What is the volume of each cube?

Cube *A*

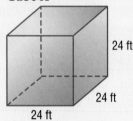

24 ft

24 ft

24 ft

Cube *B*

3

3

3

Scale Factor: $\frac{1}{8}$

Cube *A*: $V = 24 \cdot 24 \cdot 24 = 13{,}824 \text{ ft}^3$ or 512 yd^3

Cube *B*: $V = (\frac{1}{8} \cdot 24) \cdot (\frac{1}{8} \cdot 24) \cdot (\frac{1}{8} \cdot 24) =$

$\underbrace{\frac{1}{8} \cdot \frac{1}{8} \cdot \frac{1}{8}}_{\text{scale factor}} \cdot \underbrace{24 \cdot 24 \cdot 24}_{\text{dimensions}} = 27 \text{ ft}^3$ or 1 yd^3

So, the volume for Cube *B* is 1 yd^3.

The ratio of the volumes is $\frac{1}{8^3} = \frac{1}{512}$.

 Check It Out

Solve.

6 The scale factor for two similar solids is 4. What is the ratio of the volume?

7·5 Exercises

Give the scale factor.

1.

20 | 16
25 → 20

2. 14 /\ 14   →   10 /\ 10
   14   10

3. A document that measures 28 centimeters by 20 centimeters is enlarged by a scale factor of $\frac{3}{2}$. What are the dimensions of the enlarged document?

4. The scale factor of the two similar triangles is $\frac{1}{2}$. What are the dimensions of the smaller triangle?

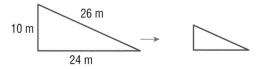

5. The scale of a map is 1 inch = 10 miles. On the map, two towns are 3.5 inches apart. How far apart are they actually?

6. The scale of a map is 1 centimeter = 5 kilometers. What length on the map would a road be that is 12.5 kilometers long?

7. The scale factor of two similar figures is 4. If the smaller rectangle has an area of 2 square feet, what is the area of the larger figure?

8. The scale factor of the two similar pentagons is $\frac{2}{3}$. Find the lengths of r, s, and t.

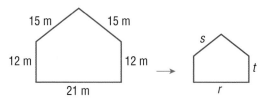

9. The scale factor of two similar solids is 5. What is the ratio of the volumes?

Measurement

You can use the problems and the list of words that follow to see what you learned in this chapter. You can find out more about a particular problem or word by referring to the topic number (*for example,* Lesson 7·2).

Problem Set

1. Which system of measurement is based on powers of ten?
   (Lesson 7·1)

2. Complete the equivalent measures. (Lesson 7·1)

   1 km = ____ m = ____ cm = ____ mm

Complete the conversions. (Lesson 7·2)

3. 32.65 km = ____ m

4. 1.28 km = ____ mm

5. $5\frac{1}{2}$ yd = ____ in.

6. 7,920 ft = ____ mi

Use the rectangle to answer Exercises 7–10.

45 in.

18 in.

7. What is the rectangle's perimeter in feet? (Lesson 7·2)

8. What is the rectangle's perimeter in yards? (Lesson 7·2)

9. What is the rectangle's area in square feet? (Lesson 7·3)

10. What is the rectangle's area in square yards? (Lesson 7·3)

Complete the conversions. (Lesson 7·3)

11. 10.04 cm^2 = ____ mm^2

12. 11 yd^2 = ____ ft^2

13. 4 yd^2 = ____ in^2

14. 7.113 m^3 = ____ cm^3

Use the rectangular prism to answer Exercises 15–17. (Lesson 7·3)

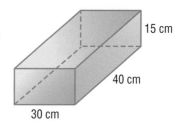

15 cm

40 cm

30 cm

15. What is the prism's volume in cubic millimeters?

16. What is the prism's volume in cubic meters?

17. If 1 inch = 2.54 centimeters, what is the approximate volume of the prism in cubic inches?

18. A recipe calls for 14 ounces of rice for 6 people. How many pounds of rice do you need to make the recipe for 36 people? (Lesson 7·4)

19. How many 175-gram packets of flour would you have to buy if you needed at least 4.4 kilograms of flour? (Lesson 7·4)

20. A 1-centimeter cube was tripled by its side lengths. What is the volume of the new cube?

A photograph that was 5 inches long by 3 inches wide was enlarged so that the new photo is 1 foot long. (Lesson 7·5)

21. What is the new width of the photo?

22. If the area of the original photo is x in^2, what is the area of the new photo in terms of x?

HotWords

Write definitions for the following words.

accuracy (Lesson 7·1)

area (Lesson 7·1)

customary system (Lesson 7·1)

distance (Lesson 7·2)

factor (Lesson 7·1)

fraction (Lesson 7·1)

length (Lesson 7·2)

metric system (Lesson 7·1)

perimeter (Lesson 7·1)

power (Lesson 7·1)

ratio (Lesson 7·6)

rectangle (Lesson 7·1)

rounding (Lesson 7·1)

scale factor (Lesson 7·6)

side (Lesson 7·1)

similar figures (Lesson 7·6)

square (Lesson 7·1)

volume (Lesson 7·3)

HotTopic 8

Tools

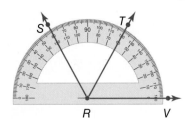

What do you know?

You can use the problems and the list of words that follow to see what you already know about this chapter. The answers to the problems are in **HotSolutions** at the back of the book, and the definitions of the words are in **HotWords** at the front of the book. You can find out more about a particular problem or word by referring to the topic number (*for example*, Lesson 8·2).

Problem Set

Use your calculator for Exercises 1–6. Round decimal answers to the nearest hundredth. (Lessons 8·1 and 8·2)

1. $55 + 8 \times 12$

2. 150% of 1,700

3. 6.4^3

4. Find the reciprocal of 5.3.

5. Find the square of 8.2.

6. Find the square root of 8.2.

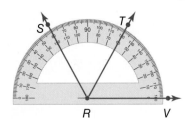

7. What is the measure of $\angle VRS$? (Lesson 8·3)

8. What is the measure of $\angle SRT$? (Lesson 8·3)

9. Does $\overrightarrow{RT}$ divide $\angle SRV$ into two equal angles? (Lesson 8·3)

10. What basic construction tool is used to draw a circle? (Lesson 8·3)

For Exercises 11–13, refer to the spreadsheet.

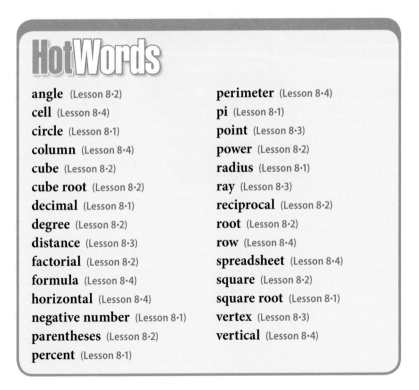

11. Name the cell with entry 40. (Lesson 8·4)

12. A formula for cell C2 is = C1 + 100. Name another formula for cell C2. (Lesson 8·4)

13. Cell D1 contains the number 2,000 and no formula. After using the command fill down, what number will be in cell D6? (Lesson 8·4)

HotWords

angle (Lesson 8·2)	**perimeter** (Lesson 8·4)
cell (Lesson 8·4)	**pi** (Lesson 8·1)
circle (Lesson 8·1)	**point** (Lesson 8·3)
column (Lesson 8·4)	**power** (Lesson 8·2)
cube (Lesson 8·2)	**radius** (Lesson 8·1)
cube root (Lesson 8·2)	**ray** (Lesson 8·3)
decimal (Lesson 8·1)	**reciprocal** (Lesson 8·2)
degree (Lesson 8·2)	**root** (Lesson 8·2)
distance (Lesson 8·3)	**row** (Lesson 8·4)
factorial (Lesson 8·2)	**spreadsheet** (Lesson 8·4)
formula (Lesson 8·4)	**square** (Lesson 8·2)
horizontal (Lesson 8·4)	**square root** (Lesson 8·1)
negative number (Lesson 8·1)	**vertex** (Lesson 8·3)
parentheses (Lesson 8·2)	**vertical** (Lesson 8·4)
percent (Lesson 8·1)	

8·1 Four-Function Calculator

People use calculators to make mathematical tasks easier. You may have seen your parents using a calculator to balance a checkbook. A calculator may not always be the fastest way to complete a mathematical task. If your answer does not need to be exact, it might be faster to estimate. Sometimes you can do the problem in your head quickly, or using a pencil and paper might be a better method. Calculators are particularly helpful for complex problems or problems that contain large numbers.

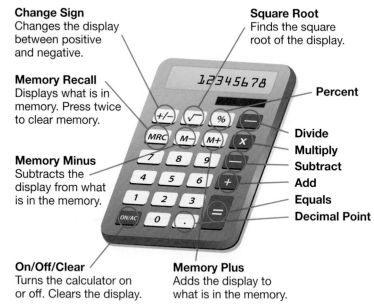

Change Sign
Changes the display between positive and negative.

Square Root
Finds the square root of the display.

Memory Recall
Displays what is in memory. Press twice to clear memory.

Percent

Divide

Multiply

Memory Minus
Subtracts the display from what is in the memory.

Subtract

Add

Equals

Decimal Point

On/Off/Clear
Turns the calculator on or off. Clears the display.

Memory Plus
Adds the display to what is in the memory.

A calculator only gives the answer to the problem you enter. Always have an estimate of the answer you expect. Then compare the calculator answer to your estimate to be sure that you entered the problem correctly.

Basic Operations

Addition, subtraction, multiplication, and division are fairly straightforward calculator functions.

Operation	Problem	Calculator Keys	Display
Addition	40 + 15.8	40 [+] 15.8 [=]	55.8
Subtraction	18 − 23	18 [−] 23 [=]	−5.
Multiplication	12.5 × 3.3	12.5 [×] 3.3 [=]	41.25
Division	8 ÷ 20	8 [÷] 20 [=]	0.4

Negative Numbers

To enter a **negative number** into your calculator, press [+/−] after you enter the number.

Problem	Calculator Keys	Display
−22 + 17	22 [+/−] [+] 17 [=]	−5.
30 − (−4.5)	30 [−] 4.5 [+/−] [=]	34.5
20 × (−8)	20 [×] 8 [+/−] [=]	−160.
−10 ÷ (−2)	10 [+/−] [÷] 2 [+/−] [=]	5.

Check It Out

Find each answer on a calculator.

1. 19.5 + 7.2
2. 31.8 − 23.9
3. 10 × (−0.5)
4. −24 ÷ 0.6

Memory

You can use the memory function for complex or multi-step problems. You operate memory with three special keys. Here is the way many calculators operate. If yours does not work this way, check the instructions that came with your calculator.

Key	Function
MRC	One press displays (recalls) what is in memory. Press twice to clear memory.
M+	Adds display to what is in memory.
M−	Subtracts display from what is in memory.

When calculator memory contains something other than zero, the display will show [M_____] along with whatever number the display currently shows. What you do on your calculator does not change memory unless you use the special memory keys.

To calculate $20 + 65 + 46 \times 5 + 50 - 5^2$ you could use the following keystrokes on your calculator:

Keystrokes	Display
MRC MRC C	0.
5 × 5 M−	M 25.
46 × 5 M+	M 230.
20 + 65 M+	M 85.
50 M+	M 50.
MRC	M 340.

The answer is 340. Notice the use of *order of operations* (p. 74).

 Check It Out

Use the memory function on your calculator to find each answer.

5 $8^4 \times 2 + (-15) \times 10 + (-21)$

6 $15^2 + 22 \times (-4) - (-80)$

Special Keys

Some calculators have keys with special functions to save time.

Key	Function
$\sqrt{x}$	Finds the **square root** of the display.
%	Changes display to the **decimal** expression of a **percent**.
π	Automatically enters **pi** to as many places as your calculator holds.

The % and π keys save you time by saving keystrokes. The √ key allows you to find square roots precisely, something difficult to do with paper and pencil. See how these keys work in the examples below.

Problem: $18 + \sqrt{225}$

Keystrokes: 18 + 225 √ =

Final display: | 33. |

If you try to take a square root of a negative number, your calculator will display an error message, such as 25 +/- √ E 5. .

There is no square root of −25, because no number times itself can give a negative number.

Problem: Find 25% of 50.

Keystrokes: 50 × 25 %

Final display: | 12.5 |

The % key only changes a percent to its decimal form. If you know how to convert percents to decimals, you probably will not use the % key much.

Problem: Find the area of a circle with radius 4.
(Use formula $A = \pi r^2$.)

Keystrokes: π × 4 × 4 =

Final display: | 50.27 |

If your calculator does not have the π key, you can use 3.14 or 3.1416 as an approximation for π.

Check It Out

Solve.

7 Without using the calculator, tell what the display would be if you entered: 10 M+ 5 × 4 + MRC = .

8 Use memory functions to find the answer to $220 - 6^2 \times (-10)$.

9 Find the square root of 484.

10 Find 35% of 250.

8·1 Exercises

Find the value of each expression, using your calculator.

1. $26.8 + 43.3$

2. $65.82 - 30.7$

3. $-16.5 - 7.46$

4. $15 \times 32 \times 10$

5. $-8 + 40 \times (-6)$

6. $75 - 19 \times 20$

7. $\sqrt{144} - 5$

8. $64 + \sqrt{256}$

9. $72 \div 16 + 12$

10. $8 \div (-40)$

Use a calculator for Exercises 11–19.

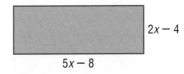

$2x - 4$

$5x - 8$

11. Find the area of the rectangle if $x = 3.5$ cm.

12. Find the perimeter if $x = 2.48$ cm.

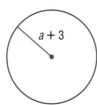

$a + 3$

13. Find the area of the circle if $a = 5$.

14. Find the circumference if $a = 1$.

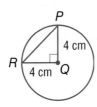

P

4 cm

R

4 cm Q

15. Find the area of $\triangle RQP$.

16. Find the perimeter of $\triangle RQP$.
    (*Remember:* $a^2 + b^2 = c^2$.)

17. Find the circumference of circle Q.

18. Find the area of circle Q.

19. Find the length of line segment RP.

8·2 Scientific Calculator

Mathematicians and scientists use scientific calculators to solve complex equations quickly and accurately. Scientific calculators vary widely; some perform a few functions and others perform many functions. Some calculators can even be programmed with functions of your choosing. The calculator below shows functions you might find on a scientific calculator.

2nd
Press to get the 2nd function for any key. 2nd functions are listed above each key.

Square Root
Finds the square root of the display.

Display

π
Automatically enters π.

On/All Clear

Clear Entry/Clear

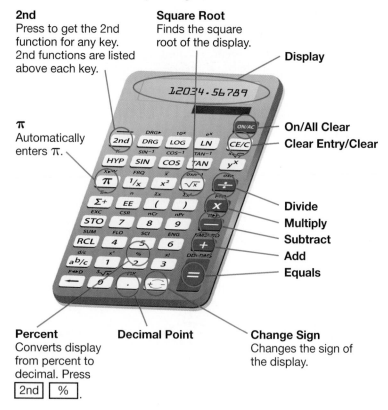

Divide

Multiply

Subtract

Add

Equals

Percent
Converts display from percent to decimal. Press 2nd % .

Decimal Point

Change Sign
Changes the sign of the display.

Frequently Used Functions

Since each scientific calculator is set up differently, your calculator may not work exactly as shown below. These keystrokes work with the calculator illustrated on page 374. Use the reference material that came with your calculator to perform similar functions. See the index to find more information about the mathematics here.

Function	Problem	Keystrokes
Cube Root $\boxed{\sqrt[3]{x}}$ Finds the cube root of the display.	$\sqrt[3]{125}$	125 $\boxed{\text{2nd}}$ $\boxed{\sqrt[3]{x}}$ $\boxed{5.}$
Cube $\boxed{x^3}$ Finds the cube of the display.	12^3	12 $\boxed{\text{2nd}}$ $\boxed{x^3}$ $\boxed{1728.}$
Factorial $\boxed{x!}$ Finds the factorial of the display.	$4!$	4 $\boxed{\text{2nd}}$ $\boxed{x!}$ $\boxed{24.}$
Fix number of **decimal places**. $\boxed{\text{FIX}}$ Rounds display to number of places you determine.	Round 2.189 to the hundredths place.	2.189 $\boxed{\text{2nd}}$ $\boxed{\text{FIX}}$ 2 $\boxed{2.19}$
Parentheses $\boxed{(}\ \boxed{)}$ Groups calculations.	$(3 + 5) \times 11$	11 $\boxed{\times}$ $\boxed{(}$ 3 $\boxed{+}$ 5 $\boxed{)}$ $\boxed{=}$ $\boxed{88.}$
Powers $\boxed{y^x}$ Finds the x power of the display.	21^4	21 $\boxed{y^x}$ 4 $\boxed{=}$ $\boxed{194481.}$
Powers of ten $\boxed{10^x}$ Raises ten to the power displayed.	10^4	4 $\boxed{\text{2nd}}$ $\boxed{10^x}$ $\boxed{10000.}$

Function	Problem	Keystrokes
Reciprocal $\boxed{1/x}$ Finds the reciprocal of the display.	Find the reciprocal of 5.	5 $\boxed{1/x}$ $\boxed{ 0.2}$
Roots $\boxed{\sqrt[x]{y}}$ Finds the x root of the display.	$\sqrt[5]{7{,}776}$	7776 $\boxed{2nd}$ $\boxed{\sqrt[x]{y}}$ 5 $\boxed{=}$ $\boxed{ 6.}$
Square $\boxed{x^2}$ Finds the square of the display.	13^2	13 $\boxed{x^2}$ $\boxed{ 169.}$

 Check It Out

Use your calculator to find the following.

1 8!

2 11^4

Use your calculator to find the following to the nearest thousandth.

3 the reciprocal of 8

4 $(9^2 - 11^4 + \sqrt[3]{1728}) \div 4$

SCIENTIFIC CALCULATOR

8·2

8·2 Exercises

Use a scientific calculator to find the following. For answers with decimals, give your answer to the nearest hundredth.

1. 45^2

2. 20^3

3. 7^4

4. 1.5^4

5. $\dfrac{33}{\pi}$

6. $7(\pi)$

7. $\dfrac{1}{4}$

8. $\dfrac{3}{\pi}$

9. $(12 + 2.6)^2 + 8$

10. $64 - (12 \div 3.5)$

11. $3! \times 5!$

12. $6! \div 4!$

13. $8! + 7!$

14. 10^{-2}

15. $\sqrt[3]{5,832}$

16. reciprocal of 20

8·3 Geometry Tools

The Ruler

To measure the dimensions of a small object, or to measure short **distances**, use a ruler.

A metric ruler

A customary ruler

To get an accurate measure, be sure that one end of the item being measured lines up with zero on your ruler.

The pencil below is measured first to the nearest tenth of a centimeter and then to the nearest eighth of an inch.

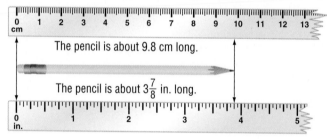

The pencil is about 9.8 cm long.

The pencil is about $3\frac{7}{8}$ in. long.

If the object or distance you are measuring is larger than your ruler, use a larger measurement tool, such as a yard or meterstick, for greater accuracy.

 Check It Out

> Use your ruler. Measure each line segment to the nearest tenth of a centimeter or the nearest eighth of an inch.
>
> **1** ─────────────────────────
>
> **2** ───────────────────────
>
> **3** ──────────────────

The Protractor

Angles are measured with a *protractor*. There are many different kinds of protractors.

The key is to find the center point of the protractor to which you align the **vertex** of the angle.

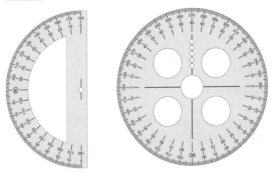

EXAMPLE   Measuring Angles With a Protractor

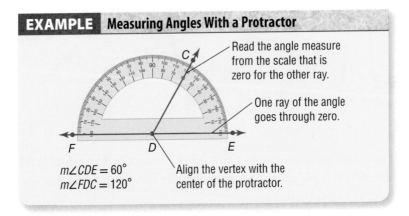

Read the angle measure from the scale that is zero for the other ray.

One ray of the angle goes through zero.

$m\angle CDE = 60°$
$m\angle FDC = 120°$

Align the vertex with the center of the protractor.

To draw an angle with a given measure, draw one **ray** first and position the center of the protractor at the endpoint. Then make a dot at the desired measure (45°, in this example).

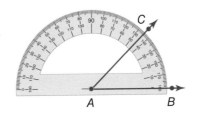

Connect A and C. Then $\angle BAC$ is a 45° angle.

Measure each angle to the nearest degree, using your
protractor.

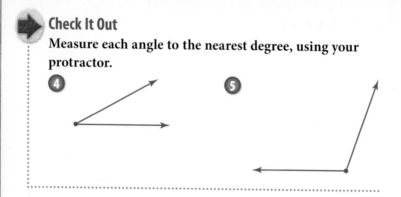

The Compass

A *compass* is used to draw circles or parts of circles, called **arcs**.
You place the fixed **point** at the center and hold it there. The
point with the pencil attached is pivoted to draw the arc or circle.

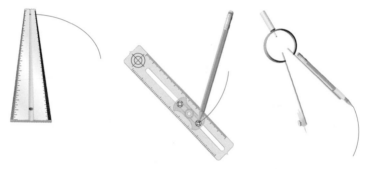

The distance between the point that is stationary (the center) and
the pencil is the radius. Some compasses allow you to set the
radius exactly.

For a review of *circles,* see page 332.

To draw a circle with a radius of $1\frac{1}{2}$ inches, set the distance between the stationary point of your compass and the pencil at $1\frac{1}{2}$ inches. Draw the circle.

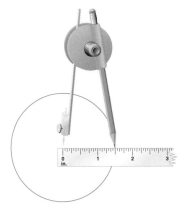

EXAMPLE   Measuring the Radius of a Circle

Measure the radius of the circle.

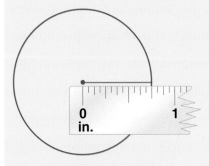

- Align the end of the ruler on the stationary with the point in the center of the circle.
- Mark where the arc of the circle falls on the ruler.
- Determine the distance.

The radius is $\frac{3}{4}$ in. or 2 cm.

Draw circles with the following radius measures.

6 radius 3 in. or 7.6 cm   **7** radius 6 cm or 2.4 in.

8 radius 4 cm or 1.6 in.   **9** radius 4 in. or 10.2 cm

Measure the radius of each circle.

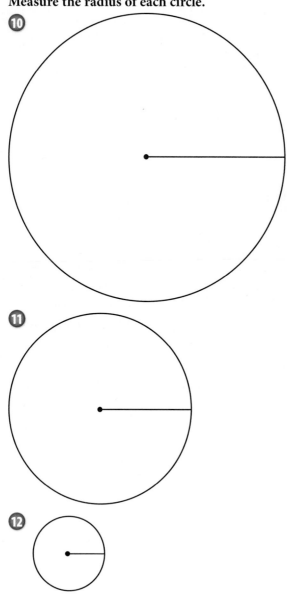

10

11

12

GEOMETRY TOOLS

8·3

Ferns, Fractals, and Branches

Have you ever looked closely at a fern frond? Did you notice that the frond is made up of smaller and smaller parts that look self-similar? That is, the smaller parts of the frond look very much like the frond itself.

In mathematics, shapes that have complex and similarly detailed structures at any level of magnification are called *fractals*. Many natural objects such as ferns display fractal-like patterns. But they are not fractals in the mathematical sense, because their complexity does not go on forever.

Use any of your geometric tools to draw a fractal pattern. It can be a design based on a geometric shape, or it can be a pattern like one found in nature.

8·3 Exercises

Using a ruler, measure the length of each side of △ABC.
Give your answer in inches or centimeters, rounded to the
nearest $\frac{1}{8}$ inch or $\frac{1}{10}$ centimeter.

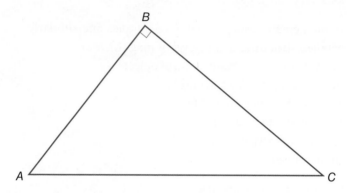

1. *AB*

2. *BC*

3. *AC*

Using a protractor, measure each angle in △ABC.

4. ∠A

5. ∠B

6. ∠C

7. If all of the sides of a triangle are equal, what do you know
   about the measure of each of the angles? Explain.

Match each tool with the function.

Tool	Function
8. ruler	A. measures angles
9. compass	B. measures distance
10. protractor	C. draws circles or arcs

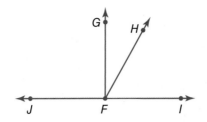

Write the measures of each angle.

11. ∠GFH   **12.** ∠HFJ

13. ∠IFJ   **14.** ∠HFI

15. ∠JFG

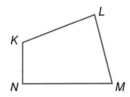

Write the measures of each angle.

16. ∠NML   **17.** ∠MLK

18. ∠KNM   **19.** ∠LKN

20. Use a protractor to copy ∠MLK.

21. Use a compass and straightedge to bisect the angle you drew for Exercise 20.

Using a ruler, protractor, and compass, copy the figures below.

22.

23.

24.

25.

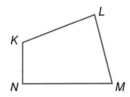

8·4 Spreadsheets

What Is a Spreadsheet?

People have used **spreadsheets** as a tool to keep track of information, such as finances, for a long time. Spreadsheets were a paper-and-pencil math tool before becoming computerized. You may be familiar with computer spreadsheet programs.

A computer spreadsheet program calculates and organizes information in **cells** within a grid. When the value of a cell is changed, all other cells dependent on that value automatically change.

Spreadsheets are organized into **rows** and **columns**. Rows are **horizontal** and are numbered. Columns are **vertical** and are named by capital letters. The cells are named according to the column and row in which they are located.

File	Edit				

Sample.xls

◇	A	B	C	D
1	1	3	1	
2	2	6	4	
3	3	9	9	
4	4	12	16	
5	5	15	25	
6				

Sheet 1 / Sheet 2 / Sheet 3

The cell A3 is in Column A, Row 3. In this spreadsheet, there is a 3 in cell A3.

Check It Out

In the spreadsheet above, what number appears in each cell?

 A1

 B3

 C4

Spreadsheet Formulas

A cell can contain a word, a number, or a formula to generate a number. A **formula** assigns a value to a cell dependent on other cells in the spreadsheet. The way the formulas are written depends on the particular spreadsheet computer software that you are using. Although you enter a formula in a cell, the value generated by the formula appears in the cell; the formula is not shown.

	A	B	C	D
1	Item	Price	Qty	Total
2	sweater	$25	2	$50
3	pants	$30	2	
4	shirt	$20	4	
5				
6				

←Express the value of the cell in relationship to other cells.

Total = Price × Qty

D2 = B2 * C2

If you change the value of a cell and a formula depends on that value, the result of the formula will change.

In the spreadsheet above, if you entered 3 sweaters instead of 2 (C2 = 3), the Total column would automatically change to $75.

Check It Out

Use the spreadsheet above. If the total is always figured the same way, write the formula for:

4 D3

5 D4

6 What is the total spent on shirts?

7 What is the total spent on pants?

Fill Down and Fill Right

Spreadsheet programs are designed to do other basic computation tasks for you. *Fill down* and *fill right* are two useful spreadsheet commands.

To use *fill down,* select a portion of a column. *Fill down* will take the top cell that has been selected and copy it into the rest of the highlighted cells. If the top cell in the selected range contains a number, such as 5, *fill down* will generate a column containing all 5s.

If the top cell of the selected range contains a formula, the *fill down* feature will automatically adjust the formula as you go from cell to cell.

The selected column is highlighted.

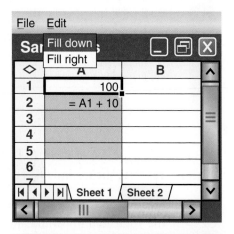

The spreadsheet fills the column and adjusts the formula.

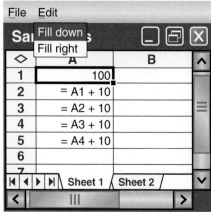

These are the values that actually appear.

Fill right works in a similar manner. It copies the contents of the leftmost cell of the selected range into each selected cell within a row.

File　Edit

Sa~~mple~~s

Fill down		
Fill right		
◇	A	B
1	100	
2	110	
3	120	
4	130	
5	140	
6		
7		

Sheet 1　Sheet 2

Row 1 is selected.

File　Edit

Sa~~mple~~s

Fill down					
Fill right					
◇	A	B	C	D	E
1	100				
2	= A1 + 10				
3	= A2 + 10				
4	= A3 + 10				
5	= A4 + 10				
6					

Sheet 1　Sheet 2

The 100 fills to the right.

File　Edit

Sa~~mple~~s

Fill down					
Fill right					
◇	A	B	C	D	E
1	100	100	100	100	100
2	= A1 + 10				
3	= A2 + 10				
4	= A3 + 10				
5	= A4 + 10				
6					

Sheet 1　Sheet 2

If you select A1 to E1 and *fill right,* you will get all 100s.

If you select A2 to E2 and *fill right,* you will "copy" the formula A1 + 10 as shown.

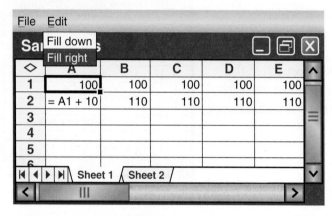

Row 2 is selected.

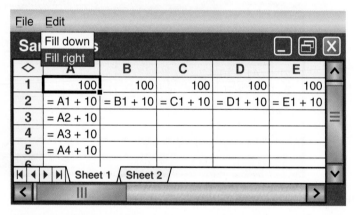

The spreadsheet fills the row and adjusts the formula.

Check It Out

Use the spreadsheet above to answer Exercises 8–10.

8 "Select" B2 to B7 and fill down. What formula will be in B6? what number?

9 "Select" A3 to E3 and fill right. What formula will be in D3? what number?

10 "Select" C2 to C5 and fill down. What formula will be in C4? what number?

Spreadsheet Graphs

You can graph from a spreadsheet. As an example, let's use a spreadsheet to compare the **perimeter** of a square to the length of a side.

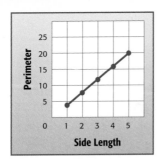

Most spreadsheets have a function that displays tables as graphs. See your spreadsheet reference for more information.

Check It Out

Use the spreadsheet above to answer Exercises 11–14.

11. What cells give the point (2, 8)?
12. What cells give the point (4, 16)?
13. What point is shown by cells A2, B2?
14. What point is shown by cells A4, B4?

8·4 Exercises

For the spreadsheet shown below, what number appears in each of the following cells?

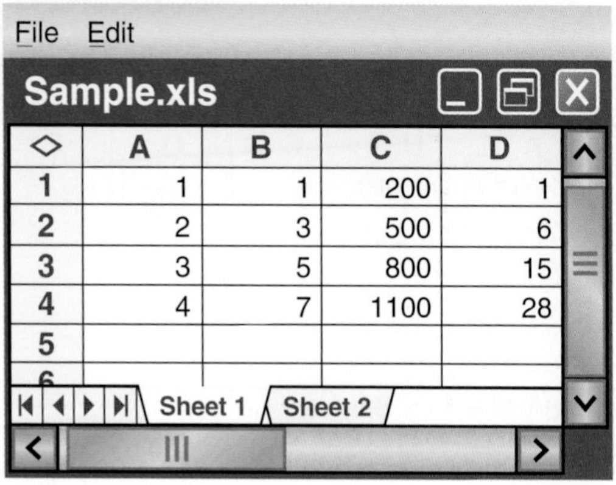

◇	A	B	C	D
1	1	1	200	1
2	2	3	500	6
3	3	5	800	15
4	4	7	1100	28
5				
6				

Sample.xls

File   Edit

Sheet 1 ⟋ Sheet 2

1. A4

2. C3

3. B1

In which cell does each number appear?

4. 1100

5. 6

6. 5

7. If the formula for cell B2 is $= B1 + 2$, what is the formula for cell B3?

8. What is the formula for cell A4?

9. If cells C5, C6, and C7 were filled in, what would the values of each of those cells be?

10. The formula for cell D2 is $= A2 * B2$. What is the formula for cell D3?

SPREADSHEETS

8·4

Use the spreadsheet below to answer Exercises 11–15.

File	Edit				
Sa	Fill down			⬜ ▣ ☒	
	Fill right				
◇	A		B		⌃
1	5		10		
2	= A1 + 6		= B1 * 2		▬
	⏮ ◀ ▶ ⏭ Sheet 1 ⟍ Sheet 2				⌄
	⟨	III		⟩	

11. If you select B2 to B6 and fill down, what formula will appear in B6?

12. Once you fill down as in Exercise 11, what numbers will appear in B3 to B5?

13. If you select A2 to D2 and fill right, what will appear in D2?

14. If you select A1 to D1 and fill right, what will appear in C1?

15. If you select A1 to A7 and fill down, what will appear in A5?

Tools

You can use the problems and the list of words that follow to see what you learned in this chapter. You can find out more about a particular problem or word by referring to the topic number (*for example,* Lesson 8-2).

Problem Set

Use your calculator for Exercises 1–10. (Lessons 8·1 and 8·2)

1. $27 + 11 \times 19$

2. 240% of 850

Round answers to the nearest tenth.

3. $70 - 18 \div (-10) + 43.1$

4. $14 + 58 \div 4.5 - 8.75$

5. 3.2^4

6. Find the reciprocal of 4.2.

7. Find the square of 4.2.

8. Find the square root of 42.25.

9. $(8 \times 10^6) \times (5 \times 10^7)$

10. $0.8 \times (55 \times 3.5)$

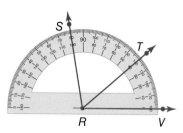

11. What is the measure of $\angle SRV$? (Lesson 8·3)

12. What is the measure of $\angle VRT$? (Lesson 8·3)

13. What is the measure of $\angle TRS$? (Lesson 8·3)

14. Does $\overrightarrow{RT}$ divide $\angle VRS$ into two equal angles? (Lesson 8·3)

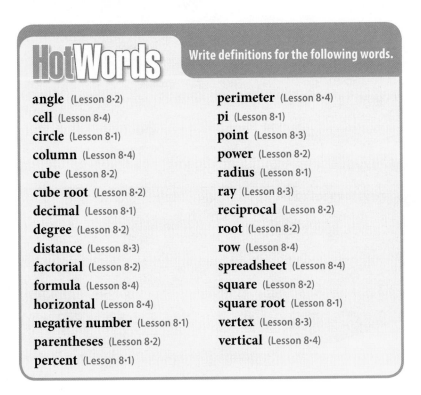

For Exercises 15–17, refer to the spreadsheet above. (Lesson 8·4)

15. Name the cell holding 9.

16. A formula for cell B2 is = B1 + 6. Name another formula for cell B2.

17. Cell D1 contains the number 240 and no formula. After using the command fill down on cells D1 through D8, what number will be in cell D8?

HotWords

Write definitions for the following words.

angle (Lesson 8·2)

cell (Lesson 8·4)

circle (Lesson 8·1)

column (Lesson 8·4)

cube (Lesson 8·2)

cube root (Lesson 8·2)

decimal (Lesson 8·1)

degree (Lesson 8·2)

distance (Lesson 8·3)

factorial (Lesson 8·2)

formula (Lesson 8·4)

horizontal (Lesson 8·4)

negative number (Lesson 8·1)

parentheses (Lesson 8·2)

percent (Lesson 8·1)

perimeter (Lesson 8·4)

pi (Lesson 8·1)

point (Lesson 8·3)

power (Lesson 8·2)

radius (Lesson 8·1)

ray (Lesson 8·3)

reciprocal (Lesson 8·2)

root (Lesson 8·2)

row (Lesson 8·4)

spreadsheet (Lesson 8·4)

square (Lesson 8·2)

square root (Lesson 8·1)

vertex (Lesson 8·3)

vertical (Lesson 8·4)

accuracy

are

estimate

slope

function

coordi

Hot Solutions

HotSolutions

Chapter ❶ Numbers and Computation

p. 68
1. 0 2. 18 3. 4,089 4. 0 5. 500 6. 1,400
7. $(4 + 7) \times 4 = 44$ 8. $20 + (16 \div 4) + 5 = 29$
9. no 10. no 11. yes 12. no 13. 5×7
14. 5×23 15. $2^2 \times 5 \times 11$ 16. 6 17. 15 18. 6
19. 15 20. 200 21. 360

p. 69
22. 12, 36, or 108 23. 8, 8 24. 14, −14 25. 11, 11
26. 20, −20 27. 7 28. −1 29. −12 30. 8 31. 0
32. 4 33. 35 34. −5 35. 6 36. 60 37. −24
38. −66 39. It is a positive integer.
40. It is a negative integer.

1•1 Properties

p. 71
1. yes 2. no 3. no 4. yes 5. 26,307 6. 199
7. 0 8. 2.4

p. 72
9. $(3 \times 2) + (3 \times 6)$ 10. $6 \times (7 + 8)$

1•2 Order of Operations

p. 74
1. 14 2. 81 3. 40 4. 26

1•3 Factors and Multiples

p. 76
1. 1, 3, 5, 15 2. 1, 2, 4, 8, 16

p. 77
3. 1, 2, 4 4. 1, 7 5. 4 6. 10

p. 78
7. 1, 3 8. 1, 2

p. 79
9. yes 10. no 11. no 12. yes

p. 80
13. yes 14. no 15. yes 16. no

p. 81
17. $2^3 \times 5$ 18. $2^2 \times 5^2$

p. 82   **19.** 2   **20.** 5   **21.** 21   **22.** 24   **23.** 120   **24.** 72

p. 83   **Darting Around** Sample answer: $(3 \times 20 + 3 \times 20 + 2 \times 20) + (3 \times 20 + 3 \times 20 + 3 \times 20) + (3 \times 20 + 3 \times 20 + 20) + (3 \times 20 + 3 \times 20 + 3 \times 20) + (50 + 50 + 3 \times 7) + (3 \times 20 + 3 \times 20 + 3 \times 20) + (2 \times 20)$

1•4 Integer Operations

p. 85   **1.** -4 **2.** $+300$

p. 86   **3.** 15, 15   **4.** 3, -3 **5.** 12, 12   **6.** >   **7.** >   **8.** <

p. 87   **9.** $-13, -2, 0, 6, 8$ **10.** $-16, -10, 6, 18, 19$

p. 88   **11.** -3 **12.** 0   **13.** -2 **14.** -4
  15. 8   **16.** -4 **17.** 4   **18.** -56

Chapter ❷ Fractions, Decimals, and Percents

p. 92   **1.** 46.901 oz   **2.** 92%   **3.** C, $\frac{18}{60}$ **4.** $\frac{29}{30}$ **5.** $1\frac{1}{7}$
  6. B, $1\frac{1}{2}$ **7.** $\frac{3}{10}$ **8.** $1\frac{6}{7}$

p. 93   **9.** $2 + 0.002$ **10.** 300.303   **11.** 14.164   **12.** 1.44
  13. 13.325   **14.** 34.8   **15.** 25%   **16.** 2   **17.** 99%
  18. 40%   **19.** 7%   **20.** 83%   **21.** $\frac{27}{100}$ **22.** $1\frac{1}{5}$
  23. 7%, 0.7, $\frac{3}{4}$

2•1 Fractions and Equivalent Fractions

p. 95   **1.** $\frac{3}{4}$ **2.** $\frac{5}{9}$ **3.** Check students' drawings for accuracy.

p. 97   **4.** Sample answers: $\frac{2}{4}, \frac{4}{8}, \frac{6}{12}$ **5.** Sample answers: $\frac{1}{2}, \frac{2}{4}, \frac{4}{8}$
  6. Sample answer: $\frac{3}{3}, \frac{7}{7}, \frac{9}{9}$

p. 98   **7.** =   **8.** =   **9.** $\neq$

p. 99   **10.** 8; $\frac{3}{8}; \frac{2}{8}$ **11.** 50; $\frac{35}{50}; \frac{21}{50}$ **12.** 10; $\frac{3}{10}; \frac{5}{10}$
  13. 36; $\frac{21}{36}; \frac{16}{36}$

p. 102 17. $3\frac{1}{6}$ 18. $1\frac{2}{3}$ 19. $1\frac{7}{9}$ 20. $7\frac{1}{8}$ 21. $\frac{35}{4}$ 22. $\frac{61}{4}$

23. $\frac{50}{3}$ 24. $\frac{59}{10}$

2•2 Comparing and Ordering Fractions

p. 105 1. $<$ 2. $>$ 3. $<$ 4. $>$ 5. $<$ 6. $>$ 7. $<$

p. 106 8. $\frac{2}{3}, \frac{5}{7}, \frac{3}{4}, \frac{5}{6}$ 9. $\frac{2}{3}, \frac{3}{4}, \frac{7}{8}, \frac{9}{10}$ 10. $\frac{1}{8}, \frac{3}{8}, \frac{5}{12}, \frac{3}{4}, \frac{5}{6}$

2•3 Addition and Subtraction of Fractions

p. 108 1. $\frac{7}{5}$ or $1\frac{2}{5}$ 2. $\frac{18}{12}$ or $1\frac{1}{2}$ 3. $\frac{5}{9}$ 4. $\frac{8}{26}$ or $\frac{4}{13}$

p. 109 5. $1\frac{1}{8}$ 6. $\frac{7}{30}$ 7. $\frac{2}{3}$

p. 110 8. 6   9. $10\frac{1}{4}$ 10. $5\frac{1}{4}$ 11. 20

p. 111 12. $7\frac{13}{18}$ 13. $10\frac{1}{3}$ 14. $7\frac{31}{36}$ 15. $7\frac{1}{2}$

p. 112 16. $3\frac{5}{6}$ 17. $2\frac{1}{2}$ 18. $5\frac{1}{4}$ 19. $1\frac{1}{2}$

2•4 Multiplication and Division of Fractions

p. 115 1. $\frac{3}{10}$ 2. $1\frac{2}{3}$ 3. $\frac{3}{6} = \frac{1}{2}$ 4. $\frac{4}{7}$ 5. 4

p. 116 6. $\frac{5}{2}$ 7. $\frac{1}{4}$ 8. $\frac{3}{7}$ 9. $\frac{7}{25}$

p. 117 10. $5\frac{1}{3}$ 11. $3\frac{1}{2}$ 12. $\frac{3}{4}$ 13. $8\frac{3}{4}$

2•5 Naming and Ordering Decimals

p. 120 1. four ones; four and four hundred eleven thousandths

2. two thousandths; thirty-two thousandths

3. four thousandths; five and forty-six ten-thousandths

4. one hundred-thousandth; three hundred forty-one hundred-thousandths   5. $<$ 6. $>$

p. 121 7. 3.0186; 3.1608; 30.618   8. 9; 9.083; 9.084; 9.381; 93.8

9. 0.6212; 0.622; 0.6612; 0.662   10. 1.54   11. 36.39

12. 8.30

HOTSOLUTIONS

2•6 Decimal Operations

2•7 Meaning of Percent

2•8 Using and Finding Percents

2•9 Fraction, Decimal, and Percent Relationships

p. 149   **1.** 60%   **2.** 30%   **3.** 90%   **4.** 208%

  5. $\frac{17}{100}$ **6.** $\frac{1}{20}$ **7.** $\frac{9}{25}$ **8.** $\frac{16}{25}$

p. 150   **9.** $\frac{129}{400}$ **10.** $\frac{191}{400}$ **11.** $\frac{985}{800}$ **12.** 27%   **13.** 0.7%

  14. 1.8%   **15.** 150%

p. 151   **16.** 0.49   **17.** 0.03   **18.** 1.8   **19.** 0.007

p. 152   **20.** 0.3   **21.** 0.875   **22.** $0.0\overline{9}$ **23.** $\frac{39}{50}$ **24.** $\frac{27}{50}$ **25.** $\frac{6}{25}$

p. 153   **26.** 0.23, 62%, $\frac{5}{7}$ **27.** 18%, $\frac{5}{9}$, 0.78   **28.** 0.25, $\frac{1}{2}$, 60%

p. 154   **The Ups and Downs of Stocks**   1%

Chapter ❸ Powers and Roots

p. 158   **1.** 3^5 **2.** n^3 **3.** 9^3 **4.** x^8 **5.** 9   **6.** 49   **7.** 16
  8. 64   **9.** 27   **10.** 64   **11.** 216   **12.** 729   **13.** 10,000
  14. 1,000,000   **15.** 10,000,000,000   **16.** 100,000,000
  17. 5   **18.** 8   **19.** 10   **20.** 9

p. 159   **21.** 5 and 6   **22.** 3 and 4   **23.** 8 and 9   **24.** 8 and 9
  25. 6.928   **26.** 7.550   **27.** 9.434   **28.** 9.899
  29. 3.6×10^7 **30.** 6×10^5 **31.** 8.09×10^{10}
  32. 5.4×10^2 **33.** 5,700,000   **34.** 1,998
  35. 700,000,000   **36.** 734,000

3•1 Powers and Exponents

p. 160   **1.** 3^4 **2.** 7^4 **3.** a^6 **4.** z^5

p. 161   **5.** 16   **6.** 64   **7.** 9   **8.** 100

p. 162   **9.** 125   **10.** 1,000   **11.** 512   **12.** 216

p. 163   **13.** 100   **14.** 1,000,000   **15.** 100,000,000   **16.** 1,000

p. 164   **When Zeros Count**   No; at the given rate, it would take almost 32 years to count to 1 billion and 10^{91} times as long to count to a googol.

3·2 Square Roots

p. 166 **1.** 3   **2.** 6   **3.** 9   **4.** 11

p. 167 **5.** between 6 and 7   **6.** between 4 and 5
7. between 2 and 3   **8.** between 9 and 10

p. 169 **9.** 1.732   **10.** 6.856   **11.** 9.274   **12.** 9.849

3·3 Scientific Notation

p. 172 **1.** 5.3×10^4 **2.** 4×10^6 **3.** 7.08×10^{10}
4. 2.634×10^7 **5.** 67,000   **6.** 289,000,000
7. 170,300   **8.** 8,520,640,000,000

Chapter ❹ Data, Statistics, and Probability

p. 176 **1.** no   **2.** biased

3.   **Number of Minutes Spent
Doing Sit-Ups**

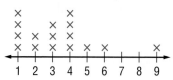

4.

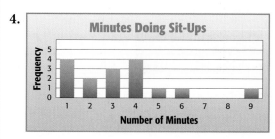

p. 177 **5.** 18 points   **6.** mean: 87.4 points; median: 88 points
7. 5   **8.** 28   **9.** 5,040   **10.** $\frac{3}{5}$ **11.** 0

4·1 Collecting Data

p. 179 1. adults registered to vote; 500
2. fish in Sunshine Lake; 200

p. 180 3. No, it is limited to people who are in that store, so they may like it best. 4. Yes, each student has the same chance of being picked.
5. Sample answer: I could assign each classmate a unique number, and then draw a certain percent of the numbers at random.

p. 181 6. Biased: A sample of people in Florida does not represent the entire population. 7. Unbiased: This is a random sample, so the sample is unbiased.

p. 182 8. It assumes that you like mystery novels.
9. It does not assume that you watch TV in the summer.
10. Do you read books?

p. 183 11. 4 12. factory 13. Science museum; more students chose that than any other possible field trip.

p. 184 **Flavor of the Week** about 25 should prefer Banana Bonanza

4·2 Displaying Data

p. 186 1. 7 and 8 letters

2.

Number of Hours Spent	0	1	2	3	4	5	6	7	8	9
Number of Students	4	6	3	3	2	0	1	1	0	2

p. 188 3. about a quarter 4. about half

5. **County Fair Entries**

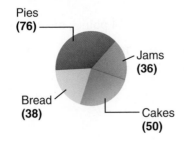

Pies (76)
Jams (36)
Cakes (50)
Bread (38)

p. 189 **6.** 7   **7.** 4

8. Number of Letters in Each Word

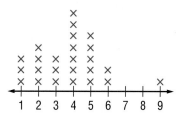

p. 190 **9.** 2002   **10.** 1999   **11.** false

p. 191 **12.** 12   **13.** 304   **14.** 22

p. 192 **15.** John Hancock Center   **16.** 280

p. 194 **17.** 16

18.

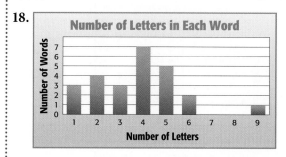

And the Winner Is . . . winning the award improved sales; weekly; bar graph

4·3 Analyzing Data

p. 197 **1.**

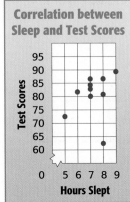

2.

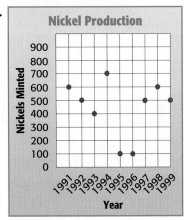

p. 199 **3.** Identifying Objects   **4.** no correlation
5. Distance Traveled

4·4 Statistics

p. 202 **1.** 11   **2.** 78   **3.** 359.5   **4.** $52

p. 203 **5.** 11   **6.** 5.5   **7.** 48   **8.** 255 ft

p. 204 **9.** 27   **10.** 4.2   **11.** 11   **12.** $0.75

p. 205 **13.** 910   **14.** 5.7   **15.** 66°   **16.** 15

4·5 Probability

p. 207 **1.** $\frac{3}{6}$ or $\frac{1}{2}$ **2.** $\frac{3}{6}$ or $\frac{1}{2}$ **3.** $\frac{3}{6}$ or $\frac{1}{2}$ **4.** $\frac{2}{6}$ or $\frac{1}{3}$

	0	1	2	3	4	5	6	7	8	9
0	00	01	02	03	04	05	06	07	08	09
1	10	11	12	13	14	15	16	17	18	19
2	20	21	22	23	24	25	26	27	28	29
3	30	31	32	33	34	35	36	37	38	39
4	40	41	42	43	44	45	46	47	48	49
5	50	51	52	53	54	55	56	57	58	59
6	60	61	62	63	64	65	66	67	68	69
7	70	71	72	73	74	75	76	77	78	79
8	80	81	82	83	84	85	86	87	88	89
9	90	91	92	93	94	95	96	97	98	99

11. $\frac{9}{100}$

13.

14.

15.

p. 226   **1.** $2x - 5 = 3$ **2.** $6(n + 2) = 20$ **3.** $6n + 2b$

4. $5n - 7$ **5.** 10 mi   **6.** $x = 6$ **7.** $y = -18$ **8.** $x = 5$

9. $y = 35$ **10.** $n = 5$ **11.** $y = -4$ **12.** $n = 4\frac{1}{2}$

13. $x = 3\frac{1}{3}$ **14.** 15 girls

p. 227   **15.**

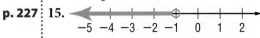

$$x < -1$$

16.

$$x \geq 3$$

17–20.

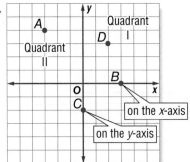

21. $-\dfrac{5}{3}$

22.

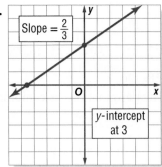

23.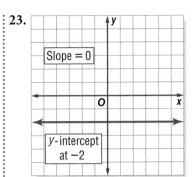

Slope = 0

y-intercept at −2

24.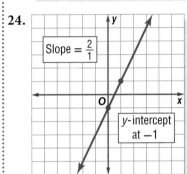

Slope = $\frac{2}{1}$

y-intercept at −1

25.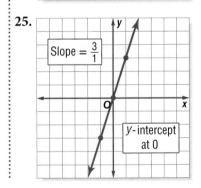

Slope = $\frac{3}{1}$

y-intercept at 0

5·1 Writing Expressions and Equations

p. 228   **1.** 2   **2.** 1   **3.** 3   **4.** 1

p. 229   **5.** $3 + x$ **6.** $n + 9$ **7.** $y + 5$ **8.** $n + 4$

p. 230   **9.** $10 - x$ **10.** $n - 7$ **11.** $y - 5$ **12.** $n - 8$

p. 231   **13.** $6x$ **14.** $4n$ **15.** $0.75y$ **16.** $10n$

p. 232   **17.** $\frac{x}{3}$ **18.** $\frac{12}{n}$ **19.** $\frac{30}{y}$ **20.** $\frac{n}{7}$

21. $5n - 8$ **22.** $2x - 4$ **23.** $2(n - 10)$

24. $3x - 6 = 6$ **25.** $5 + \frac{n}{4} = 10$ **26.** $3y - 2 = 25$

Orphaned Whale Rescued $2,378 + 25d = 9,000$

5·2 Simplifying Expressions

1. no **2.** yes **3.** no **4.** yes

5. $3 + 2x$ **6.** $5n$ **7.** $3y + 6$ **8.** 6×5
9. $3 + (7 + 10)$ **10.** $4 \times (2 \times 7)$ **11.** $4x + (3y + 5)$
12. $(4 \times 5)n$

13. $3(50 - 2) = 144$ **14.** $6(100 - 1) = 594$
15. $5(100 - 15) = 425$ **16.** $7(200 - 1) = 1,393$

17. $10x + 4$ **18.** $18n - 12$ **19.** $-6y + 4$ **20.** $8x - 10$

21. $7(x + 3)$ **22.** $3(4n - 3)$ **23.** $10(c + 3)$
24. $5(2b + 5)$

25. $13x$ **26.** $4y$ **27.** $10n$ **28.** $-4a$ **29.** $5y + 6z$
30. $7x - 15$ **31.** $14a - 24$ **32.** $20n - 4$

5·3 Evaluating Expressions and Formulas

1. 14 **2.** 2 **3.** 15 **4.** 14

5. 20°C **6.** −20.6°C **7.** 0°C

8. 110°F **9.** 9°F **10.** 204.8°F **11.** 87.8°F
12. 30 mi **13.** 2,275 km **14.** 330 mi **15.** 12 ft

5·4 Solving Linear Equations

1. true, false, false **2.** false, true, false
3. false, true, false **4.** false, false, true

5. yes **6.** no **7.** no **8.** yes

9. $x + 3 = 15$ **10.** $x - 3 = 9$ **11.** $3x = 36$
12. $\frac{x}{3} = 4$ **13.** −4 **14.** x **15.** 35 **16.** $-10y$

5•5 Ratio and Proportion

5•6 Inequalities

5•7 Graphing on the Coordinate Plane

p. 270 9–12.

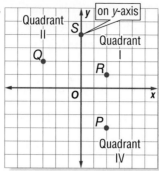

p. 272 13–16.

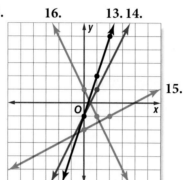

p. 273 17–20.

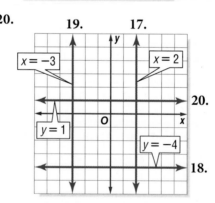

5•8 Slope and Intercept

p. 276 1. $\frac{3}{2}$ 2. $-\frac{3}{4}$

p. 277 3. -1 4. $\frac{3}{2}$ 5. $-\frac{1}{2}$ 6. 4

p. 278 7. 0   8. no slope   9. no slope   10. 0

p. 279 11. 0   12. 4

p. 280 13–16.

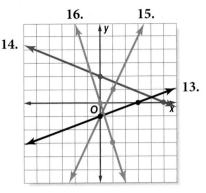

14. 16. 15. 13.

17. slope $= -3$, y-intercept at 2

18. slope $= \frac{1}{4}$, y-intercept at -2

19. slope $= -\frac{2}{3}$, y-intercept at 0

20. slope $= 6$, y-intercept at -5

Chapter **6** Geometry

p. 284 1. 62° 2. 28° 3. H, X, O 4. 14 cm 5. 60 ft²
6. 56 in² 7. 1,256 m² 8. 125 in³ 9. 80 cm³

p. 285 10. 16π m² 11. 5 in. 12. 12 in.

6•1 Naming and Classifying Angles and Triangles

p. 287 1. $\overleftrightarrow{KL}$, $\overleftrightarrow{LK}$ 2. K

p. 288 3. N 4. ∠KNL or ∠LNK, ∠LNM or ∠MNL, ∠KNM or ∠MNK 5. 60° 6. 85° 7. 145°

p. 289 8. 45°; acute angle 9. 180°; straight angle
10. 135°; obtuse angle

p. 290 11. supplementary 12. vertical

p. 291 13. congruent 14. complementary

p. 293 15. $m\angle L = 30°$ 16. $m\angle C = 45°$ 17. $m\angle E = 60°$

6·2 Polygons and Polyhedrons

1. Sample answers: *ABCD; ADCB; BCDA; BADC; CDAB; CBAD; DABC; DCBA* **2.** 360° **3.** 65°

4. no; yes; no; yes; no **5.** Yes; it has four sides that are the same length and opposite sides are parallel.

6. yes; hexagon **7.** yes; quadrilateral **8.** no

A Tangram Zoo

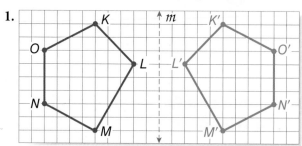

9. 900° **10.** 120°

11. pentagonal prism **12.** square pyramid

6·3 Symmetry and Transformations

1.

Flip It, Slide It, Turn It Messages Can math be fun?

2. no **3.** yes; one **4.** yes; two **5.** yes; two

6. 270° **7.** 180°

8. yes; 4 units right and 1 unit up **9.** no
10. yes; 4 units left and 4 units up

6·4 Perimeter

1. $16\frac{1}{2}$ in.   **2.** 56 m   **3.** 4 ft   **4.** 21 m

5. 30 ft   **6.** 36 in.   **7.** 16 cm

The Pentagon 924 ft

8. 36 m   **9.** 30 cm   **10.** 8 in.

6·5 Area

1. about 46 m^2

2. 324 in^2 **3.** 169 m^2 **4.** 48 cm^2 **5.** 11 ft   **6.** 40 ft^2

7. 75 cm^2 **8.** 120 in^2

9. 25 m^2 **10.** 17 cm^2 **11.** 19.5 in^2

6·6 Surface Area

1. 250 in^2 **2.** 22 m^2

3. 900 m^2 **4.** B   **5.** ≈ 408.2 in^2

6·7 Volume

1. 5 m^3 **2.** 3 cm^3 **3.** 36 cm^3

4. 420 ft^3 **5.** 48 in^3

6. $\approx 62,800$ ft^3 **7.** $\approx 1,570$ cm^3

6·8 Circles

1. 13 m   **2.** 0.5 cm   **3.** 32 in.   **4.** 5 ft   **5.** 12 m

6. 28π ft   **7.** 21π cm   **8.** ≈ 45.8 m   **9.** ≈ 113 cm
10. ≈ 6.5 in.

11. 169π cm^2 **12.** ≈ 346 ft^2 **13.** 64π cm^2

Now, That's a Pizza! $\approx 2,507$ ft^2

6•9 Pythagorean Theorem

p. 339   **1.** 9, 16, 25   **2.** Area A + Area B = Area C

p. 340   **3.** 8 m   **4.** 20 cm

Chapter ⑦ Measurement

p. 344   **1.** Sample answers: gram, kilogram; ounce, pound, ton
      2. Sample answers: millimeter, centimeter, meter,
      kilometer; inch, foot, yard, mile   **3.** 4,500
      4. 8,400   **5.** 144   **6.** 0.25   **7.** 2,280   **8.** 63
      9. 1,296   **10.** 1,900,000   **11.** 360 in.   **12.** 5,184 in^2

p. 345   **13.** 5,000,000 mm^3 **14.** 0.005 m^3 **15.** about 305 in^2
      16. 3.75 lb   **17.** 12 packets   **18.** 6 in.   **19.** $\frac{4}{9}$

7•1 Systems of Measurement

p. 347   **1.** metric   **2.** customary

7•2 Length and Distance

p. 350   **1.** 2.2   **2.** 5   **3.** 108   **4.** 500

p. 351   **5.** 78.7 cm   **6.** 70 yd   **7.** A   **8.** B

7•3 Area, Volume, and Capacity

p. 353   **1.** 63 ft^2 **2.** 40,000 cm^2

p. 354   **3.** 0.125 m^3 **4.** 10 yd^3

p. 355   **5.** the quart   **6.** the pint   **7.** the liter

7•4 Mass and Weight

p. 357   **1.** 4,200   **2.** 0.64   **3.** 88.2

p. 358   **Poor SID**   No, the mass is always the same.

7·5 Size and Scale

p. 359   **1.** Q and U are similar. R and T are similar.

p. 360   **2.** $\dfrac{5}{2}$ **3.** $\dfrac{3}{4}$

p. 361   **4.** $\dfrac{25}{16}$ **5.** 100 ft^2

p. 362   **6.** 343 times   **7.** $\dfrac{64}{1}$

Chapter ❽ Tools

p. 366   **1.** 151   **2.** 2,550   **3.** 262.14   **4.** 0.19   **5.** 67.24   **6.** 2.86
       7. 120°   **8.** 60°   **9.** yes   **10.** compass

p. 367   **11.** B2   **12.** A2 × B2   **13.** 2,000

8·1 Four-Function Calculator

p. 369   **1.** 26.7   **2.** 7.9   **3.** −5   **4.** −40

p. 370   **5.** 8,021   **6.** 217

p. 371   **The Mystery of Memory** Sample answer: Immediate memory is used for entering the numbers; short-term memory is used for remembering the running total; and long-term memory is used for remembering the rules and arithmetic facts.

p. 372   **7.** 30   **8.** 580   **9.** 22   **10.** 87.5

8·2 Scientific Calculator

p. 376   **1.** 40,320   **2.** 14,641   **3.** 0.125   **4.** −3,637

8·3 Geometry Tools

p. 378   **1.** 7.6 cm or 3 in.   **2.** 7 cm or $2\dfrac{3}{4}$ in.   **3.** 2 in. or 5.1 cm

p. 380   **4.** 28°   **5.** 110°

p. 382   **6–9.** Measure radius to check.   **10.** $1\dfrac{1}{2}$ in. or 4 cm
       11. 1 in. or 2.5 cm   **12.** $\dfrac{3}{8}$ in. or 1 cm

8·4 Spreadsheets

p. 386　**1.** 1　**2.** 9　**3.** 16

p. 387　**4.** B3 × C3　**5.** B4 × C4　**6.** $80　**7.** $60

p. 390　**8.** = B5 + 10; 150　**9.** = D2 + 10; 120
10. = C3 + 10; 130

p. 391　**11.** A3, B3　**12.** A5, B5　**13.** (1, 4)　**14.** (3, 12)

Index

involving division, 232
involving multiplication, 231
involving subtraction, 230
with two operations, 232–233

F

Faces of polyhedrons, 303
Factorial function key, 375
Factorial notation, 215
Factoring out common factors, 240
Factorization, prime, *see* Prime factorization
Factors of numbers, 76–78
canceling, 115
common, 77, 78
in customary system of measurement, 347
greatest common factor (GCF), 77
negative, Distributive Property of Addition and Multiplication with, 239
Fahrenheit degrees, 245
False statements, 248
Flip It, Slide It, Turn It messages, 307
Flips, 306
Fluid ounces, 355
Formulas
for area of a circle, 326, 335
for circumference of a circle, 333
for converting temperatures, 245
for distance traveled, 246
evaluating, 245
for probability of simple event, 207, 209
for probability of target event, 220
for simple interest, 145
for slope of a line, 275, 276
on spreadsheets, 387
for volume
of a cylinder, 330
of a prism, 329
Four-function calculators, 368–372
Fractals, 383
Fraction method for finding percents, 136
Fractions, 94
addition of
with like denominators, 108
with unlike denominators, 109
comparing, 104
converting
to decimals, 152
decimals to, 152
mixed numbers percents to, 149
to percents, 148
percents to, 149

in customary system of measurement, 347
decimals and, 151–152
division of, 117
equivalent, 96–98, 104
expressing probability as, 209
improper, *see* Improper fractions
like, 98
multiplication of, 114–115
shortcuts for, 115
names for one, 96
naming, 94–95
ordering, 106
percents and, 148
subtraction of
with like denominators, 108
with unlike denominators, 109
unlike, 98
writing, in simplest form, 100
Frequency table, 193

G

GCF, *see* Greatest common factor (GCF)
Geometry tools, 378–382
Googol, 164
Graphing
on coordinate plane, 268–273
equations of lines, 271–273
equations with two variables, 271–273
inequalities, 264–265
lines using slope and y-intercept, 279
Graphs
bar, 191–192
circle, 187
double-bar, 192
histograms, 193
line, 189–190
Greater than (>), 264
Greater than or equal to (≥), 264
Greatest common factor (GCF), 77, 100
prime factorization in finding, 81

H

Heptagon, 299
Hexagon, 299
angles of, 301
Histograms, 192, 193
Horizontal lines, 273
slopes of, 278
Horizontal number line, 268
Horizontal rows in spreadsheets, 386
Hypotenuse of triangle, 338, 339

O

P

Writing
  equations, 233
  expressions
    involving addition, 229
    involving division, 232
    involving multiplication, 231
    involving subtraction, 230
    involving two operations, 232–233
  fractions in simplest form, 100
  inequalities, 265
  multiplication using exponents, 160
  ordered pairs, 269
  very large numbers in scientific
    notation, 117

X-axis, 268
X-coordinate, 269

Y-axis, 268
Y-coordinate, 269
Y-intercept, 278
  graphing lines using slopes and, 279

Zero (or Identity) Property of Addition, 71
Zero Property of Multiplication, 71
Zeros
  in powers of ten, 163, 164
  products of decimals with, 126

Photo Credits

All coins photographed by United States Mint.
002–003 Dennis Welsh/Getty Images; **066–067** CORBIS;
083 Jeffrey Coolidge/Getty Images; **088** Robert Lehmann/Alamy;
130 Yann Layma/Getty Images; **154** CORBIS; **164** Getty Images;
184 Maximilian Weinzierl/Alamy; **194** Erik S. Lesser/AP Photo;
218 MazerStock; **234** Debra Behr/Alamy; **251** PunchStock;
258 James Porto/Getty Images; **314** Getty Images; **336** Brand X
Pictures; **358** Scott Camazine/Alamy; **371** Ryan McVay/Getty
Images; **383** Adam Burton/Alamy; **396–397** Masterfile.